STUDIES IN IMPERIALISM

General editors: Andrew S. Thompson and Alan Lester
Founding editor: John M. MacKenzie

When the 'Studies in Imperialism' series was founded
by Professor John M. MacKenzie more than thirty years
ago, emphasis was laid upon the conviction that
'imperialism as a cultural phenomenon had as significant
an effect on the dominant as on the subordinate societies'.
With well over a hundred titles now published, this
remains the prime concern of the series. Cross-disciplinary
work has indeed appeared covering the full spectrum of
cultural phenomena, as well as examining aspects of
gender and sex, frontiers and law, science and the
environment, language and literature, migration and
patriotic societies, and much else. Moreover, the series has
always wished to present comparative work on European
and American imperialism, and particularly welcomes the
submission of books in these areas. The fascination with
imperialism, in all its aspects, shows no sign of abating,
and this series will continue to lead the way in
encouraging the widest possible range of studies in the
field. 'Studies in Imperialism' is fully organic in its
development, always seeking to be at the cutting edge,
responding to the latest interests of scholars and the needs
of this ever-expanding area of scholarship.

Building the French empire, 1600–1800

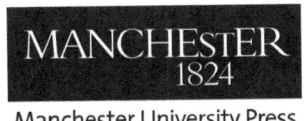

Manchester University Press

Building the French empire, 1600–1800

COLONIALISM AND MATERIAL CULTURE

Benjamin Steiner

MANCHESTER UNIVERSITY PRESS

The right of Benjamin Steiner to be identified as the author of this work has been asserted by them in accordance with the Copyright, Designs and Patents Act 1988.

Published by MANCHESTER UNIVERSITY PRESS
Oxford Road, Manchester M13 9PL

www.manchesteruniversitypress.co.uk

British Library Cataloguing-in-Publication Data
A catalogue record for this book is available from the British Library

ISBN   978 1 5261 4323 5   hardback
ISBN   978 1 5261 6701 9   paperback

First published 2020
Paperback published 2022

The publisher has no responsibility for the persistence or accuracy of URLs for any external or third-party internet websites referred to in this book, and does not guarantee that any content on such websites is, or will remain, accurate or appropriate.

Typeset
by New Best-set Typesetters Ltd

To Mi Anh

CONTENTS

ILLUSTRATIONS

ACKNOWLEDGEMENTS

This book has been written far away from the places I have chosen to study from a perspective as a historian interested in the plurality and differences in the early modern world. Except for a journey to Martinique I undertook together with my wife in September 2017, I have not travelled to the places and regions of the former early modern French empire. I regret this very much, but I have soothed myself that this book could rather serve as an incentive to go to these places and explore them with the knowledge I gathered from the archival sources. From my experience in Martinique, however, I have learned that other things than written sources matter to historians, too. Besides the archeological evidence that I tried to collect from the respective literature one becomes aware of the materiality of the remains of the old building structures; at least I did when I visited the ruins of Château Dubuc on the Caravelle peninsula near the town of La Trinité on Martinique. I was also surprised to see how the places I knew from the sources were situated in relation to each other, how long one had to travel to reach them, and how the topographical situation really looked like. Today it is easy to drive around the magnificent landscape of this island. But it was different for the inhabitants of the seventeenth century to go from one place to the other, not to speak of the problems transport and logistics for the large building projects faced with the island's topography.

But I have not trained as an archaeologist or as an ethnographer who has the expertise to analyse and present the objects and locales in present times. I tried to get along with the written sources of the archives, dig through administrative correspondence, lists, and tables of building material and workers, the journals and memos of engineers, and tried to learn how a place like Martinique actually got included into the dominion of the French empire. What I discovered was that empire building has not been an achievement solely by European ingenuity, inventiveness, and audacity, but has relied on other non-European people, their actions and knowledge as well. I found this discovery important enough to describe it in detail and give it support with historical evidence. This objective, to show how a colonial empire was created by a diverse and plural group of people, remains the main purpose of my book. I hope the reader will excuse my lack of local knowledge of the present day and nonetheless gain insight from this book about this early period of colonialism.

As for myself, I have been inspired to continue my quest of learning more about the differences and entanglements in the early modern world, and will try to continue travelling to the places I described in this book: the settlements on the banks of the Senegal River, the cities of Puducherry (Pondichéry) and Karikal in India, Atlantic Canada and its partly reconstructed town of Louisbourg, the Antillean islands of Guadeloupe and Saint-Kitts (Saint-Christophe), and the towns and cities of Haiti (formerly Saint-Domingue). Perhaps the readers of my book already have travelled to all of these places and know more about them than me. But I might give these readers more reason to see the interconnections and similiarties that bind these places to each other. This is a legacy of the early modern empire that was not held together by a dominant power in the metropolitan centre of France, but by the diverse community of people that inhabited these places before and after Europeans arrived at their shores and eventually left them again.

This book could not have been written without the generous support of the Max-Weber-Kolleg in Erfurt and the Kulturwissenschaftliche Kolleg in Konstanz over a period of three years. Both institutions provided me with not only the time and funding, but also the opportunity to discuss with colleagues the ideas and concept underlying this work. I received valuable advice, especially with regard to the question of the concept of empire, from scholars in a wide range of disciplines that congregate at these great centres of erudition. Hopefully, I have been able to translate their insights and guidance into a plausible narrative.

In particular, I would like to thank my colleagues from Erfurt, where I started working on this project. Martin Mulsow, Markus Vinzent, Jutta Vinzent, Riccarda Suitner, Susanne Rau, Knud Haakonssen, Jörg Rüpke, and Martin Fuchs are some of the scholars that took great interest in my research. Their recommendations, comments, and critiques have helped to refine the argument and contributed to its clarity. I would like to express my special gratitude to Giovanni Tarantino, who encouraged me to think more about the connection between empire and emotion.

A conference in Erfurt on 'Global Intellectual History' that I organized in 2016 together with Martin Mulsow brought together scholars from many fields to uncover forgotten connections in global history with regard to the exchange, entanglement, and itineraries of knowledge and ideas. Locating knowledge and ideas in different areas of the world reveals the connectivity of immaterial and material agents. I therefore owe a debt of gratitude to the contributors to the proceedings of this conference, among others, Hans Medick, Carlo Ginzburg, Kapil Raj, Dominic Sachsenmeier, Sebastian Conrad, Anna Akasoy, and Paola

ACKNOWLEDGEMENTS

Molino, from whom I gained much inspiration for the writing of this present volume.

Finishing the manuscript on the banks of Lake Constance afforded me the pleasure of writing amid beautiful scenery and the support at every step along the way of my colleagues at the Kolleg in Konstanz, who are wonderful people as well as erudite scholars. I would like to mention them to convey my appreciation for the time taken to discuss the project, their critique of the material, and their encouragement to me to complete the book. My thanks go to Robert Kramm, Paula Bialski, Eva Blome, Ulrich Bröckling, David Collins, Pim Griffioen, Susanne Lüdemann, Madeleine Reeves, Albert Schirrmeister, Rudolf Schlögl, Gabriela Signori, Martial Straub, Michael Stürner, Christina Zuber, Zuzanna Dziuban, Marcus Twellmann, Aleida Assmann, and Albrecht Koschorke, as well as Christina Thoma and Fred Girod.

Building the French empire owes its original idea of using material sources and sources on materiality for the history of the formation of empire to my previous collaboration with my colleagues from the University of Munich. They form a group of empire historians who are driven by questions similar to my own. Their work on the Spanish and Dutch colonial empires were of value for comparative purposes and enduringly influence my ideas on the subject at hand. I therefore extend my gratitude to Arndt Brendecke, Susanne Friedrich, and Vitus Huber.

Finally, I would like to thank all those students who have been patient enough to listen to my lectures on empires in Erfurt, Frankfurt, Munich, and Fribourg. Often their questions and comments went straight to the heart of the central problems of the history of empire formation, why they were established, how they were kept up as functioning polities, and who the people were, particularly those of non-European descent, that made them a built and lived reality.

Introduction: Building the French empire

I have always stood up for God and King and I will pour the last drop of my blood for my people, as well as for my country and for Spain.[1]

These words were written 25 August 1792 by Jean-François Papillon, former African slave, insurgent, and 'vice-admiral' of the black rebel force on Saint-Domingue, the western part of Hispaniola held by the French. He wanted to assure his ally Don Joaquin García, captain-general of Santo-Domingo, the eastern part of the island belonging to Spain, of his allegiance to the monarchical order of the Ancien Régime.[2] This reference to a European king by one of his formerly enslaved subjects may be surprising, but it was by no means an exception in the ranks of black military officers on Saint-Domingue at this time.

It was only one year before that the so-called Haitian Revolution had unfolded in the aftermath of a slave rebellion in 1791 led by Dutty Boukman, a high priest of voodoo and leader of the Maroon slaves. Although this rebellion was intertwined with the colonial aftermath of the French Revolution that followed the events of 14 July 1789, its leaders did not renounce Louis XVI, King of France, as their formal head of state. In fact, on 24 August, one week after the imprisonment of the King, the main figures of the revolt, Toussaint Louverture and Georges Biassou, together with Jean-François, called for all officers to gather with their troops in Grande-Rivière for a formal parade. Biassou proclaimed himself viceroy and promised 'to maintain order while awaiting instructions from the King our master, whose rights I hope to support, with the help of the Lord, until it pleases him to send us his own established laws'.[3]

This parade in honour of Louis XVI as it took place in the northern part of Saint-Domingue could just as well have been happening, as Philippe Girard remarks, in Versailles.[4] The white banner of the Bourbon

[1]

regime flew over the camp – easily imagined as a place where the military order of the rebel forces imitated that of the French royal army. Ranks, titles, and names of the former slaves remained French: Biassou, for example, called himself 'General of the Army of the King'; Jean-François kept the custom of omitting his last name, usually that of his former owner; and the future leader of the Black Republic, Toussaint Louverture, did not stop stressing his loyalty to the Bourbon family even during the period when the revolutionary commission of Sonthonax and Poverel established the Jacobin republican system on Saint-Domingue.[5]

How could it be that the link between the Antillean slave colony and the royal government in Versailles was so persistent that it even outlasted the abolition of slavery in the colonies and of the French monarchy itself? The question goes to the heart of the subject of this book, which examines the French empire and how this global polity can be understood in terms of its function, stability, and coherence. As there are several claims that deny the formal existence of a French empire before 1804, when Napoleon Bonaparte became emperor of France, the task at hand does not pose itself as obvious. But considering the bond that existed between Louis XVI and his subjects in the colonies (for example, Toussaint Louverture, who, despite his socially inferior status, maintained his allegiance), the existence of some sort of empire that maintained this relationship is not inconceivable.

The French empire, however, was not like others. Its realization did not follow a plan that emerged in the course of French exploration to places overseas. It materialized in ways that are not clearly drawn out in the written administrative documents of the national archives. The empire that I am going to reconstruct here had different roots reaching beyond the confines of a nationally formed France. This realm was indeed conceived and enacted by many people from Europe, but also from Africa, Asia, and the Americas. They were responsible for achieving not the ideological, but the material construction of empire, by improving places, landscapes, and even spaces to become cities, plantations, and territories where political, social, and cultural interaction could take place.

In this study I would like to follow these builders of empire at the construction site itself: the engineers, artisans, experts, workers, and slaves – all those builders who were responsible for the establishment of material constructs that formed the backbone of the early modern French empire. If anything, they were agents in the effort to create imperial spaces far away from their original homes in order to establish economically and socially sustainable environments. The construction of empire was a consequence of diverse agents following different, sometimes converging and sometimes diverging, interests. Empire, therefore, is here understood as a historiographical conception that does not adhere to the usual centre–periphery model of colonial and imperial historiography. It rather describes the plurality of agents and places that became integral parts of a realm that was only formally attached to a centre of administration. The building and buildings of empire, the processes and resulting monuments, represented an important binding force of this global cultural formation.

In exploring this proposition, the book draws from several recent developments in historiography, including the study of material expressions of culture and emotion. While its wide geographical scope requires that the history of empire be viewed from a global vantage point, this needs to be combined with a microhistorical perspective to avoid the presumption a default globality of historical events or a Eurocentric sense of global space that itself is a construct of the imperial past.

Microhistories of empire

Empires are elusive. The term 'empire' implies a certain ambiguity about power relations: while determined by dominant central rule, empires are also characterized by a certain weakness of government control over large distances. While this is true for most empires in world history, it is particularly noticeable in the case of the French colonial realm in the early modern period. Some historians have therefore argued that states like France, which maintained outposts, settlements, or even territories outside of Europe, had a rather ineffective hold over their colonial possessions.[6] It could even be argued that it is inappropriate to call such a rule imperial or to speak of an empire at all. Centre–periphery models of the French empire have been tested very critically. Recent studies questioned the ability of absolutist governments to extend their rule over large global distances.[7]

On the other hand, the idea of a French empire still remains as an *idée fixe* in historiography, not only of colonial or imperial history, but also in post-colonial studies and within the newly emerging field of global history.[8] Even the early, barely traceable tracks of colonial engagement in the sixteenth century have been framed in relation to a master narrative of European expansion, which eventually led to the establishment of a dominant empire in the nineteenth and twentieth centuries.

If the predominant approach to colonial and imperial history has been to understand the historical space within which empires are formed as having a centre governing the periphery, the alternative would be to try to describe the early formation of empire in relative spatial terms.

Some historians have invoked a pericentric approach, concentrating not on a single centre, but on its alleged frontiers, borders, or fringes.[9] Guiseppe Marcocci has remarked that such an approach would free historians 'from the old narrative of European expansion, whose empires are now explored more and more from their limits'.[10] He suggests viewing empires as a 'set of relational dynamics, often supported by outsiders and groups operating on the fringes'.[11] The relativistic character of early modern spaces of empire allows the metaphorical description of global passages as 'corridors for travel and trade', in which 'imaginary water channels traced peculiar transregional connection, much like wormholes of space in astronomy'.[12]

If we adopt a modified theory of spaces of empire that disregards the distinction of centre and periphery and instead concentrates on microspaces of empire, then it makes sense to follow Francesca Trivellato's suggestion to allow the potential of a microhistorical approach to benefit global history.[13] Both global and local dimensions are interconnected within imperial locations and deserve a fitting historiographical approach. The original ambition of Italian microhistorians has been to elucidate the broad trends in history through studying 'the exceptional normal' (Edoardo Grendi) in close detail, to look out for 'clues' (spie) (Carlo Ginzburg), details that are incongruent with standard narratives, and to arrive at implications that transcend the local.[14] The microhistorical approach's focus on synchronic relations, as opposed to the diachronic relations of macrohistories, unveils the different dimensions of context that surround the lives of individuals, the existence of buildings, or even the emergence of ideas. Different spheres of connection intersect and overlap in those areas chosen for microhistorical scrutiny.

Therefore empire historians require a 'large microhistory' (Emma Rothschild) that provides us with enough information and data about the extensions of microspatial confinements.[15] In following personal names, things, or ideas across multiple records, new ways to connect microhistories emerge. An example of this might be to follow certain individuals of various provenience who lived in 'global contexts'.[16] Alternatively, a microhistorical perspective could be applied to the history of slaves in order to understand the fates of individuals within the global context of the history of slavery.[17] Accounts of global microhistory offer, as Randy Sparks has shown in the case of a slaving entrepôt in Anomabu on the coast of Guinea, insights into African agency in creating the Atlantic slave trade system, beyond them being simply reduced to slaves.[18]

Microhistorical study serves as a useful lens for comparative history. Through a differentiated understanding of local conditions, of the notions of mixed agency and materiality, the hidden connections will emerge to form a picture of the larger structure of empire. In what follows, these conditions will be studied within the French empire, alternating between examples from the Antilles that provide a microhistorical perspective, and places in India, West Africa, and North America for a broader view.

Colonial project making

The new materialism in cultural studies offers an opportunity to correct the silencing of agency by focussing on the role of material objects in imperial history.[19] The work of Chandra Mukerji on the building of the gardens of Versailles and of the Canal du Midi in southern France exemplifies how the history of engineering not only reflects the technological achievements by a multitude of actors and agents, but also the social and political consequences these projects had for the establishment of modern societies and polities.[20] In extending this scholarship to the French colonial realm, the present study aims to uncover the contributions of many forgotten people – engineers, artisans, slaves, etc. – to the building of empire.

Studying large building projects in the context of empire building in the early modern period begs a question: were these undertakings already an expression of a plan or an intention to establish an imperial presence in the world? The promotion of large projects over long distances is generally considered to be a trademark of modern colonial regimes, thought of by some scholars as typical for the relatively short period from the 1870s to the 1960s. Some authors, like James C. Scott, have addressed the colonial legacy of European states' undertaking of ambitious projects in the nineteenth and twentieth centuries, and have criticized the construction of 'white elephants' as part of a modernization ideology that eventually caused more unintended damage to colonized societies than it did them good.[21]

A comparable scheme of modernization in the early modern period can be found only in a few cases in the age of European expansion. Some early modern colonial projects, however, deserve mentioning, not because they automatically fit into the modernization narrative, but because they deliver insights that reconfigure the variable constellations within which colonization took place. One of those cases is the drainage of the basin of Mexico, an enormous project resulting in the construction of the Desagüe de Huehuetoca after nearly two centuries. The study by Vera Candiana on the environmental transformation of Mexico emphasizes the interaction between Spanish and indigenous agencies that sheds light on a more complicated relationship between colonizers and colonized people.[22]

Other case studies on large colonial projects have been less focused on indigenous agency. For the planning and building of Dutch colonial towns overseas, Ron van Oers mostly sees Dutch traditions and schemes to apply the 'mother concept' of such architecture theorists as Simon Stevin at work.[23] Interestingly, van Oers follows a narrative similar to that of high modernization that gives precedence to the value of the Dutch urban heritage. This heritage eventually contributed to the diffusion of an 'organizational pattern or structure at ground level' that then persisted, as exemplified with the cities of Jakarta (Batavia), New York (New Amsterdam), and Cape Town.[24]

This unflinchingly positive narrative of successful colonial planning, however, does not surface in the relatively small body of literature that exists for building projects in the French colonial realm. On the contrary: historians of the French empire have pointed out the frailty of colonial rule and concluded that France could barely maintain law and order in its settlements, let alone execute the successful planning and building of infrastructure. Shannon Lee Dawdy even diagnosed a form of 'rogue colonialism' for the case of New Orleans.[25] Even though she rejects the idea of crediting either the success or failure of colonial projects to a strong imperial centre, she develops the concept of a 'rogue empire', in which a 'creole variant of colonialism' mediated through a heterogeneous group of archetypical figures ('the engineer, the creole, and the rogue') that created colonialism.[26] Even the eighteenth-century foundation of Louisbourg, a fortified town on the coast of Atlantic Canada, was, in the eyes of A. J. B. Johnston, not so much the product of central planning, but of a local, everyday struggle between colonial elites and settlers.[27] In Johnston's book on Louisbourg the leading protagonists of the colony are also engineers, like, for example, François Verville, the military engineer whose geometrical plans were often compromised by individual disorderly house builders.

The microhistory of individual colonial building projects as a means to portray the plurality of imperial agency has been the subject of earlier studies concerning the construction of fortresses along the West African coast. Arnold W. Lawrence, inspired by the research of his famous brother, Thomas E. Lawrence, on European castles in the Middle East, studied the construction of these buildings to uncover the participation of Africans as highly skilled workers and artisans.[28] The example of the building of Fort Anomabu in today's Ghana is a particularly illustrative episode of this shared agency in the construction of the monuments of European trade colonialism in Africa. In a detailed case study on Cape Coast Castle and its surroundings, William St Clair develops a picture of the construction site where most of the building material had to be imported from England and other places. Bricks, for example, had to be shipped from London, Bristol, and Liverpool; lime from Portland, and special stones to filter water came from Barbados. Only some essential resources like seashells had to be acquired from local African suppliers.[29]

This line of argument is also followed by Randy J. Sparks's contribution in his book on Anomabu, where he gives an account of those local professionals engaged in the construction of the fortress: 'linguists, gold-takers, farmers, market-sellers, slave traders, tomboys, craftsmen, artisans, and labourers'.[30] Of particular interest, regarding the formation of a local imperial style, is the construction of a private building called Castle Brew right outside the fortress. The structure was intended to 'awe the Fante', the local indigenous people, as well as the 'British inside the fort, and European traders who came there to do business':

> The Georgian British Palladian building, constructed of brick and stone, boasted arches, arcades, and an elegant black-and-white marble walk leading from the rear courtyard. A double staircase led up to the veranda and the entrance that most visitors would have used to reach the elegant reception hall. One departure from the Palladian style was an exterior staircase leading to the second story, the one concession to local building styles and one of several construction details the castle shared with the neighbouring fort.[31]

Lawrence, St Clair, Sparks, and others show the complex nature of imperial agency in the context of the building of large houses, fortresses, and other public works to improve the infrastructure of colonial trading posts, production colonies, and settlements.[32] While the human agency, however, was plural and variegated, the style and the materiality gave colonial places a certain homogeneity, a common style, and, finally, an imperial identity that consisted of both colonial as well as Creole elements.

Creole identity is usually understood as a state that already involves colonial elements besides its indigenous or non-colonial character. But in practice, colonial identity remained an opposite social formation to Creole identities. In the French Antilles this became apparent in the social stratification of three major groups, comprising, firstly, African slaves, secondly, free Eurafricans, the Creole, 'Câpre' or Métis (also: 'Métif') population, and, thirdly, European settlers, the 'white' population, mostly distinguished in 'small' and 'great whites' (*petits et grands blancs*). Historically and conceptually, the Creole or mestizio logics can be conceived as a phenomenon of mixing in the broadest sense. Alternative concepts such as that of hybridity have been criticized before since they imply a binary logic of two fixed and homogenous identities.[33]

Style and emotions

A central theme of large colonial building projects was style. Applying this aesthetic category to a colonial or even Creole context is not unproblematic. French architectural style, like the *architecture classique* that radiated from such archetypical buildings as those by Louis Le Vau and others in Vaux-le-Vicomte and in Versailles, the Orangerie by Jules Hardouin-Mansart, and the Pavillon d'Aurore by Charles Perrault in Sceaux, or the Corderie Royale by François Blondel in Rochefort, far beyond the borders of metropolitan France, exercised a strong influence on French colonial style. But the components of this classical colonial style were accompanied by indigenous architectural elements that sometimes appeared in contrast to colonial style or intermingled in a Creole fashion to form a new aesthetic.

Even if interest in style received more attention in recent debates among art historians, the importance of style in this study goes beyond the question of its precise definition or its academic status within the history of art or architecture.[34] What is of more interest here is how the style and style elements of large representative buildings affected the people in their environment. Therefore, style is understood in this instance as an expression of certain intentions, reflecting imperial, colonial, class, or economic interests. On the other hand, however, style takes shape in the perception of those who are confronted with its appearance. Style, therefore, underlies a certain state of emotional self-awareness.[35]

This state of emotional awareness, represented by the affective capacity of buildings and their style and elements of style, can be described as the imperial binding force that gave coherence to the plurality of people and interests. Buildings and their style helped to create 'emotional communities' not unlike those that Barbara Rosenwein described for the monastic orders in the early Middle Ages.[36] Departing from William Reddy's premise that emotions can be learned and that emotional talk and gestures – Reddy calls them 'emotives' – can alter the state of people and their environment by establishing 'emotional regimes' and 'emotional refuges',[37] the concept of Rosenwein's community leaves more room for the plurality and diversity of different emotions within a given space.[38] In this model, smaller circles are distributed unevenly, not entirely concentric, throughout a larger circle that 'is the overarching community, tied together by fundamental assumptions, values, goals, feeling rules, and accepted modes of expression', while the smaller ones 'represent subordinate emotional communities, partaking in the larger one and revealing its possibilities and its limitations'.[39]

[8]

Colonial spaces with their built environment can figure as 'affective spaces' (Andreas Reckwitz) that constituted the conditions where all colonial, Creole, and indigenous practices took place.[40] On an island like Saint-Domingue there were, for example, both places of 'suffering' and 'refuge', that is, the plantations and the maroon settlements in the interior. The towns on the island figured as spaces where diverse affects, embedded in indigenous, Creole, and colonial practices, competed with each other and contributed to the formation of an imperial identity. Individual constructions were, in this respect, carriers of emotions and affects, too, and were indeed affective buildings that interacted with their inhabitants and the people that lived nearby. Benno Gammerl has underlined this in stating that certain 'spatial settings shape the way in which diverging emotional patterns and practices come to the fore';[41] thus architecture displays not only style in the sense of layout, structure, and ornament, but also spatially defined 'emotional style'.[42] Emotional styles varied from one place to the other and could be re-appropriated by different groups of people in order to articulate or fashion their particular group or individual identity.

The advantage of concentrating on emotions within an imperial space is that this captures the plurality of affective interaction within different spheres (or circles) while upholding the idea of a larger sphere that encompasses this plurality. This larger sphere constitutes an emotional community that was constructed over a long period of time. But from the beginning it already posed as a spatially confined entity that conditioned the historical emergence of *empire* as a political formation of cultural (and emotional) coherence.

Architecture and engineering

It is not the intention of this book to provide a complete account of the history of building in the French colonial realm. Neither is it about the history of colonial architecture, and it does not deal with the urban development of colonial towns and cities. This has been done to a much more thorough extent, at least with regard to the French Atlantic, by Laurent Vidal, Émile d'Orgeix, Gilles-Antoine Langlois, and Gauvin Alexander Bailey's recent book.[43] Notably, Bailey's study on colonial architecture is accompanied by a databank of the 'Colonial Architecture Project', which he launched with his students from Queen's University in Kingston in 2017, and which provides thousands of photographs of colonial architecture not only from the French, but also the British, Spanish, Portuguese, Danish, and Dutch empires.[44] Bailey's work, therefore, is invaluable for further comparative studies on colonial and imperial architecture and urban history in the early modern period and provides a valuable collection of the existing material sites and remains. Painstakingly, Bailey has also been able to follow the builders of colonial architecture through notarial archives in order to bring to life not only the architects, contractors, masons, carpenters, and joiners trained in France, white civilians, but also aboriginal builders of the Americas and West Africa and the contributions made by African slaves and *gens de couleurs*.[45]

In addition to Bailey's empirical findings, the question of how buildings and the process of building contributed to the political and social formation of empire remains a topic to be addressed in the present study. The French empire itself is a political enigma and paradoxical in the sense of having been 'big, strong, and influential' as well as 'poorly organized, hopelessly unrealistic, plagued by mad utopian schemes, underfunded, and underpinned by a brutal and volatile institution of agricultural slave labour that proved to be one of the main sources of its undoing'.[46] This paradoxical diagnosis has been supported in the literature on the French empire again and again without offering more than the impression of a rather dysfunctional political construct that mostly lacked the features of a more functional (i.e. modern) colonial empire.

The terms 'engineering' and 'building' are, for the purpose at hand, more preferable to 'architecture' in order not only to give an account of architectural history that had been forgotten by specialists until Bailey uncovered its rich heritage, but also to analyse the function of building in the sense of what I call the material construction of empire. 'Engineering' and 'building' are conceived in this sense as a process involving not only the planning, organizing, building, and repairing of colonial and imperial architecture, but also state and empire building. The expressions 'engineering empire' or 'empire building' aim to capture the double nature of the material and aesthetic construction of buildings as well as of the political construction of imperial identity, emotional spaces, and administrative coherence.

The original French term *'ingénieur'* carried, alongside its technical meaning for experts of military and civil construction works, a certain moral connotation of being capable of perfection, accomplishment, and even wisdom.[47] In this sense, the combination of 'engineering' and 'empire' alludes to the social and political constructive effort to build an empire and can thus be understood as engineering of a polity.[48] The term 'engineering' also conveys more of the continuity of building than architecture, which rather alludes to a process with a beginning, consisting in planning, and an end, when a structure is actually finished. Engineering, in a certain way, never ends. It has to deal with interruption, deterioration, and repair, which sometimes can take as long as a building is used or is finally left to total decay.

In cultural anthropology, a trend called 'repair studies' sees technology as a function that reacts to the 'modern infrastructure ideal' with what is being called 'broken world thinking'.[49] Rather than asserting innovation, development, or design this conceptual approach favours terms such as 'breakdown', 'dissolution', and 'change' when it comes to describe new media or technology. For my purpose of laying out the process of building *and* repairing in the early modern colonial world, it also makes sense to argue in a similar way that infrastructure or complex architectural structures are not a result of modernization in the sense of building monuments that stand forever. In favouring the activities of repair, another aspect of early modern building and engineering comes to light, which shows that broken structures and things do not necessarily mean failure or defeat of an alleged plan, but stand for stability and the maintenance of order.

Structure of this book

The French empire was administratively frail, but left behind physical and non-physical traces of a strong political formation, nonetheless. Engineers and builders of empire created several colonial spaces across the globe that served as enclosures for distinct emotional communities. Large building projects stand out in their capacity to have an affective influence on the people in- and outside these colonial spaces. But the problem that follows from this is the manner in which the respective places and their communities shared more than merely their coincidental and nominal French ownership. Saint-Domingue's emotional community certainly displayed different characteristics from those of Canada or places in India, West Africa, the Mascarenes, and even the neighbouring island colonies of Martinique and Guadeloupe. Are all these possessions of France to be distinguished and studied as singular cases of colonial rule, or is it possible to detect elements of an imperial identity that was shared by these colonies? Or was there, in fact, an empire not in the sense that it was inherently 'French', but as a cultural formation that 'belonged' to the people that built, participated, lived, and felt with it?

In order to grasp the ambivalent character of empire, singular in every place but also part of a shared identity, the methodology of choice is microhistory, fittingly described in the Italian tradition by Trivellato as 'global history on a small scale'.[50] Thus each of the following chapters is dedicated to a different aspect of the French colonial realm, serving as a close-up case study highlighting the everyday practices of the engineering of empire. The microhistory of the planning, financing, managing, and execution of the large building projects in the Caribbean (followed in ever-closer detail), India, North America, and West Africa serves as a spyglass revealing the local structure of global entanglements. The chapters interlink perspectives from place to place and unveil the global connections that must be established in the process of local empire building.

The intertwined structure of the book, giving brief accounts of several different locations in the French empire, opens up a broader perspective. Rather than providing a systematic comparison, these sketches invite the reader to gain further insights into the interweaving of certain styles, practices, and identities. This approach is not mere comparativism, rather the intention is to avoid too great a focus on diversity whilst common features are ignored. The examples of Pondichéry in India, Fort Royal on Martinique, Saint-Louis and Gorée Island in Senegambia, Louisbourg in Canada, as well as Saint-Domingue, now Haiti, were chosen to offer scope for the study of the French empire, but also to show that local agency indeed made a marked difference in the imperial realm. Each place produced different imperial spaces that forced the French to adapt to local conditions in order to pursue the idea of a global empire.

A study of the history of the building of the French empire unfolds a manifold tableau of actors, things, and ideas interacting with one another on a local and a global level. The immobile character of monumental buildings favours the definition of the empire by local and stable elements as a physical reality that could be seen, sensed, and felt, foremost by those who lived, worked, and suffered nearby. Thus, the study of empire engineering poses the fundamental question as to whether the colonial presence of Europeans unintentionally brought about the emergence of communities, existing in their own right and providing people with a new identity, some having lived there for generations, others arriving as migrants, settlers, or slaves, and eventually forming societies and establishing their national identity on the constructed foundation of their colonial past.

This book will seek to explore the materiality of colonial project making, the style and emotion buildings conveyed, and the work of architects, engineers, and artisans in the following chapters. The first chapter looks into large building projects in the so-called seugneurial period on the Antillean islands of Martinique, Guadeloupe, and Saint-Christophe in the seventeenth century. It focuses on the French military engineer and architect François Blondel and his plan for the fortification system on the islands. Castles, fortresses, and batteries were not only projected for the defence against other European powers, but also to appropriate the space as a territorially enclosed colony. This practice of colonial enclosure is contrasted in the second chapter with the case of Pondichéry as an imperial city within the Mughal state system in India. The challenge to establish a stronghold within a more powerful polity made it necessary for the French administration under the governor Dupleix to include local brokers and intermediaries within the project of empire building. That led not only to more collaboration with Indian and Tamil experts in logistics, construction, and engineering, but also to gradually establish the colonial city as political entitity within the state system on the subcontinent.

Having outlined these two ways of adaptation to and appropriation of local conditions at the far sides of the French empire, the third chapter turns to the construction sites in the Antilles with the building of Fort Royal on Martinique as a case in point. While the French had to struggle to maintain this fortress due to a considerable lack of resources, workforce, and financial funds, they relied heavily on a group of free African and Creole entrepreneurs who, besides the enslaved labourers from the plantations, supplied the construction site with skilled workers, artisans, and independent subcontractors to produce and deliver building materials. Bringing to light collaboration between Africans and Europeans, the fourth chapter considers the case of the French settlements in the Senegal region. In Saint-Louis and on the island of Gorée the situation was different from the Antilles (and other French colonies) in regard to the way elements of style intermixed while being dominated by local African ornamental and formal architectural attributes. The 'mixed society' of the Senegambia region created its very own colonial style and laid the basis for an emotional community within the coastal settlements where Africans and Europeans could both identify with the building projects.

Returning to the Caribbean case, the fifth chapter presents a detailed study of engineers, builders, and building materials illustrating the functions of a network of actors that used local traditional techniques and European expertise in order to maintain the building sites on the islands. Not only did Africans and Europeans become visible as social factions belonging to this group of builders, but the Carib indigenous population also contributed with techniques, expertise, and style to the emergence of an assembled material manisfestation of a Creole identity. The picture of an 'ideal' French colonial city had been different when more resources from metropolitan France were available. The contrasting case of Louisbourg in Atlantic Canada is presented in the sixth chapter and elaborates how the strategic importance of this place helped to create a model of a segregated community within the walls of the fortified city. French settlers established an emotional regime in Louisbourg that maintained strong ties to France while excluding the Mi'kmaq, the indigenous population of the Île Royale (today Cape Breton Islands).

The last chapter shows that the model of the ideal colonial city did not always work in practice, even when colonial administrators planned building and infrastructure in employing considerable human and material resources. In contrasting the aesthetic functions of two styles of 'affective buildings' it becomes apparent that a classical colonial architecture (fortresses, government buildings, fountains) competed with a Creole variant of architecture (mansions, town houses, cabins) to represent the collective identity of the pluralist society of the islands. The case of the public water supply system for the city of Port-au-Prince in the second half of the eighteenth century illustrates a building project that involved soldiers, indentured servants, African slaves, and paid artisans, mostly of African descent, but at the same time aimed to represent the superiority of French colonial culture. The episode serves as a prelude to the Haitian revolution that indicates a continuity of appropriating expertise, techniques, styles, and material from each other. Thus the early modern practice of empire building continued in the form of nation bulding in this first republic of freed slaves.

Written and material sources

Sooner or later every attempt to write history from a global perspective is confronted with the problem of the asymmetric diffusion of written sources. Historians have thought long and hard about how to circumvent this problem in some way, in order to ensure that those people who were not able to write, or for whom nobody felt obliged to write, also have a voice in history.[51] But written sources do exist within European archives that testify to the role and deeds of such people, namely, the rather dry registries of lists and tables, and maps and plans, that provide information on the progress of the work itself.[52] Although one is not able to glean from these the individual biographies of each worker, artisan, or transporter, the lists do sometimes give information about names, gender, age, qualifications, and in some cases a character profile.

Correspondence, memos, and journals, on the other hand, privilege the agency of higher officials and those engineers responsible for the oversight of construction works.[53] The group of engineers, administrators, and those who directed the building projects in a top-down fashion are necessarily more prominent in these sources. This does not mean, however, that the building itself was conducted according to an ideal hierarchy. Material and archaeological evidence enables the introduction of other actors onto the scene who introduced new building materials and techniques, as well as distinct styles. This evidence, however, is not always retrieved by excavation or forensic analysis. Few places in the former colonies in Africa, Asia, or the Caribbean Sea have been the subject of archaeological surveys.[54] The only extensive research conducted is on sites in North America – in Québec, for example.[55] Only recently has archaeology in West Africa received more funding, and it has already produced interesting results.[56]

Currencies and measures

Some explanation of currencies and measures is required. The *livre tournois*, mostly abbreviated as tt, originally struck in Tours (hence the name), was generally used as currency from 1667 until 1795. One livre could be divided into 240 *denier*s or into 20 *sous*. In the islands of the French Antilles, New France, and the Mascarenes the colonial livre was issued in paper form and linked to the metropolitan currency at the rate of 1½ colonial livres to 1 livre tournois. In India, most of the expenses were calculated in local currency, chiefly in Indo-French rupees and pagodas. One pagoda, a gold coin equivalent to 32 rupees, could be exchanged for 8 livres 10 sous.

In engineering and construction, the chief unit for linear and cubic measurement was the *toise*. Until 1812, 1 toise was 6 feet (about 1.949 metres). A square toise was often used for measuring the masonry, equalling 3.8 square metres, also standardized under Colbert in 1667. A measure of distance and area particularly used in the Antilles was the *pas géométrique* that equalled 0.61 metres in the metric system. The *reserve des cinquante pas du roi* established by the Colbert administration designated a coastal area in Martinique, Guadeloupe, and later in Réunion that was exempt from sale to private plantations in order to reserve the littoral space of the islands for fortifications and the building of residences for fishermen, masons, and carpenters without the necessary means to afford housing elsewhere on the islands.[57]

Notes

1 Jean-François [Papillon] to Don Joaquin García, 25 August 1792, quoted in a letter by García to Don Andres de Heredia, Commander of the Northern frontier, 31 August 1792, in Antonio del Monte y Tejada, *Historia de Santo Domingo* (Santo Domingo: Garcia Herman, 1890), vol. 3, p. xii: 'yo siempre me llevado por nuestro dios y el rey y derramaré hasta la ultima gota de mi sangre por mi pueblo, igualmente por mi patria y la de Espana.'
2 For the Afro-French-Spanish discursive relations on Hispaniola during the Black revolt, see Maria Cristina Fumagalli, *On the Edge: Writing the Border between Haiti and the Dominican Republic* (Liverpool: Liverpool University Press, 2015), pp. 107–28.

3   Biassou to the Abbé de la Haye, 24 August 1792 (CARAN D/XXV/5, d. 48), quoted in Jeremy D. Popkin, *A Concise History of the Haitian Revolution* (Malden, MA/ Oxford: Wiley-Blackwell, 2012), p. 50.

4   Philippe Girard, *Toussaint Louverture: A Revolutionary Life* (New York: Basic Books, 2016), pp. 121–2.

5   See Popkin, *Concise History*, p. 59.

6   Eric Hinderaker, *Elusive Empires: Constructing Colonialism in the Ohio Valley, 1673–1800* (Cambridge: Cambridge University Press, 2008).

7   Kenneth Banks, *Chasing Empire across the Sea: Communication and the State in the French Atlantic, 1713–1763* (Montreal/London/Ithaca: McGill-Queen's University Press, 2003); Ronald J. Dale, *The Fall of New France: How the French Lost a North American Empire, 1754–1763* (Toronto: Lorimer, 2004).

8   For the early colonial historiography, see Pierre Pluchon, *Histoire de la colonisation française*, vol. 1: Le Premier empire colonial. Des origines à la Restauration (Paris: Fayard, 1991); Philip P. Boucher, *The Shaping of the French Colonial Empire: Bio-bibliography of the Careers of Richelieu, Fouquet, and Colbert* (New York: Garland Press, 1985); Herbert Ingram Priestley, *France Overseas through the Old Régime: A Study of European Expansion* (New York/London: Appleton-Century, 1939); for accounts that include the concept of empire in non-Eurocentric narratives Richard White, *The Middle Ground: Indians, Empires, and Republics in the Great Lakes Region, 1650–1815* (Cambridge: Cambridge University Press, 1991) is the seminal publication.

9   For a larger conceptual overview, cf. Michael Doyle, *Empires* (Ithaca, NY/London: Cornell University Press, 1984), p. 220; also Ann Laura Stoler and Frederick Cooper, 'Between Metropole and Colony: Rethinking a Research Agenda', in Ann Laura Stoler and Frederick Cooper (eds), *Tensions of Empire: Colonial Cultures in a Bourgeois World* (Berkeley, CA: University of California Press, 1997), pp. 1–58; Ann Laura Stoler, 'Imperial Formations and the Opacities of Rule', in Craig J. Calhoun, Frederick Cooper, and Kevin W. Moore (eds), *Lessons of Empire: Imperial Histories and American Power* (New York: New Press, 2006), pp. 48–60; Alison Games, *The Web of Empire: English Cosmopolitans in an Age of Expansion, 1560–1660* (Oxford: Oxford University Press, 2008). A particular polycentric approach to the Spanish empire is argued by Pedro Cardim, Tamar Herzog, Jose Javier Ruiz Ibáñez, and Gaetano Sabatini (eds), *Polycentric Monarchies: How Did Early Modern Spain and Portugal Achieve and Maintain a Global Hegemony?* (Brighton/Portland: Sussex Academic Press, 2012).

10 Giuseppe Marcocci, 'Too Much to Rule: States and Empires across the Early Modern World', *Journal of Early Modern History* 20 (2016), 515.

11 Ibid., 522.

12 Lauren Benton, *A Search of Sovereignty: Law and Geography in European Empires, 1400–1900* (Cambridge: Cambridge University Press, 2010), p. 108.

13 Francesca Trivellato, 'Is There a Future for Italian Microhistory in the Age of Global History?', *California Italian Studies* 2 (2011), https://escholarship.org/uc/item/0z94n9hq, retrieved 22 January 2020; see also Romain Bertrand and Guillaume Calafat, 'La microhistoire globale: affaire(s) à suivres', *Annales: Histoire, Sciences Sociales* 73 (2018), 3–19.

14 Edoardo Grendi, 'Micro-analisi e storia sociale', *Quaderni storici* 35 (1977), 512; Carlo Ginzburg, *Miti emblemi spie: morfologia e storia* (Turin: Einaudi, 1986); published in English as *Clues, Myths, and the Historical Method*, trans. John Tedeschi and Anne Tedeschi (Baltimore, MD: Johns Hopkins University Press, 1989); Giovanni Levi, *L'eredità immateriale: carriera di un esorcista nel Piemonte del Seicento* (Turin: Einaudi, 1985).

15 Emma Rothschild, *The Inner Life of Empires: An Eighteenth-Century History* (Princeton, NJ: Princeton University Press, 2011), p. 269.

16 Tonio Andrade, 'A Chinese Farmer, Two Black Boys, and a Warlord: Towards a Global Microhistory', *The Journal of World History* 21:4 (2011), 573–91.

17 Michael Zeuske, *Sklaven und Sklaverei in den Welten des Atlantiks 1400–1940: Umrisse, Anfänge, Akteure, Vergleichsfelder und Bibliographien* (Münster: Lit Verlag, 2006).

18 Randy J. Sparks, *The Two Princes of Calabar: An Eighteenth-Century Atlantic Odyssey* (Cambridge, MA: Harvard University Press, 2004).

19 Chandra Mukerji, *From Graven Images: Patterns of Modern Materialism* (New York: Columbia University Press, 1983); Bruno Latour, *Reassembling the Social: An Introduction into Actor-Network-Theory* (Oxford: Oxford University Press, 2005); Diana Coole and Samantha Frost (eds), *New Materialisms: Ontology, Agency, and Politics* (Durham/London: Duke University Press, 2010); Andreas Folkers, 'Was ist neu am neuen Materialismus? Von der Praxis zum Ereignis', in Tobias Goll, Daniel Keil, and Thomas Telios (eds), *Critical Matter: Diskussionen eines neuen Materialismus* (Münster: Edition Assemblage, 2013), pp. 17–35.

20 Chandra Mukerji, *Impossible Engineering: Technology and Territoriality on the Canal du Midi* (Princeton, NJ: Princeton University Press, 2009); Chandra Mukerji, *Territorial Ambitions and the Gardens of Versailles* (Cambridge: Cambridge University Press, 1997). See also Chandra Mukerji, 'Cartography, Entrepreneurialism, and Power in the Reign of Louis XIV: The Case of the Canal du Midi', in Pamela H. Smith ad Paula Findlen (eds), *Merchants and Marvels: Commerce, Science, and Art in Early Modern Europe* (New York: Routledge, 2002), pp. 248–76; Chandra Mukerji, 'Stewardship Politics and the Control of Wild Weather: Levees, Seawalls, and State Building in 17th-Century France', *Social Studies of Science* 37:1 (2007), 127–33.

21 James C. Scott, *Seeing Like a State: How Certain Schemes to Improve the Human Condition Have Failed* (New Haven, CT/London: Yale University Press, 1998); Dirk van Laak, *Weiße Elefanten: Anspruch und Scheitern technischer Großprojekte im 20. Jahrhundert* (Stuttgart: DVA, 1999).

22 Vera S. Candiani, *Dreaming of Dry Land: Environmental Transformation in Colonial Mexico City* (Stanford, CA: Stanford University Press, 2014).

23 Ron van Oers, *Dutch Town Planning Overseas During VOC and WIC Rule* (Zutphen: Walburg, 2000), p. 168.

24 Ibid., p. 172.

25 Shannon Lee Dawdy, *Building the Devil's Empire: French Colonial New Orleans* (Chicago, IL: University of Chicago Press, 2009), p. 5. See also Cécile Vidal, 'Africains et Européens au pays des Illinois durant la période française (1699–1765)', *French Colonial History* 3 (2003), 51–68.

26 Dawdy, *Building the Devil's Empire*, pp. 12 and 21.

27 A. J. B. Johnston, *Control and Order in French Colonial Louisbourg, 1713–1758* (East Lansing, MI: Michigan State University Press, 2001).

28 A. W. Lawrence, *Trade Castles and Forts of West Africa* (London: Jonathan Cape, 1963), pp. 53–5.

29 William St Clair, *The Grand Slave Emporium: Cape Coast Castle and the British Slave Trade* (London: Profile Books, 2007), pp. 188–9.

30 Randy J. Sparks, *Where the Negroes Are Masters: An African Port in the Era of the Slave Trade* (Cambridge, MA/London: Harvard University Press, 2014), p. 18.

31 Ibid., p. 76.

32 See also Emily Mann, 'To Build and Fortify: Defensive Architecture in the Early Atlantic Colonies', in Daniel Maudlin and Bernard L. Herman (eds), *Building the British Atlantic World: Spaces, Places and Material Culture, 1600–1850* (Chapel Hill, NC: University of North Carolina Press, 2016), pp. 31–52; Emily Mann, 'Viewed from a Distance: Eighteenth-Century Images of Fortifications on the Coast of West Africa', in Kohn Kwadwo Osei-Tutu and Victoria Smith (eds), *Shadows of Empire in West Africa: New Perspectives on European Fortifications* (London: Palgrave Macmillan, 2017), pp. 107–36.

33 Jean-Loup Amselle, *Mestizo Logics: Anthropology of Identity in Africa and Elsewhere*, trans. Claudia Royal (Stanford, CA: Stanford University Press, 1998). 'Câpre' is a French term predominantly used in the Antilles for 'light-skin' mulatto. See Ginette

Curry, *"Toubab La!" Literary Representations of Mixed-Race Characters in the African Diaspora* (Newcastle: Cambridge Scholars Publishing, 2007), pp. 141–2.

34 See, for example, Irving Lavin, 'Introduction', in Erwin Panofsky, *Three Essays on Style*, ed. Irving Lavin (Cambridge, MA: The MIT Press, 1995), pp. 3–4; James Elkins, 'Style', in *Grove Art Online: Oxford Art Online*, www.oxfordartonline.com/subscriber/article/grove/art/T082129, retrieved 19 August 2017; Jas Elsner, 'Style', in Robert S. Nelson and Richard Shiff (eds), *Critical Terms for Art History*, 2nd ed. (Chicago, IL/London: University of Chicago Press, 2003), pp. 98–109; Svetlana Alpers, 'Style Is What You Make It: The Visual Arts Once Again', in Berel Lang (ed.), *The Concept of Style*, rev. and exp. ed. (Ithaca, NY/London: Cornell University Press, 1987), pp. 138–162.

35 See Lavin, 'Introduction', p. 9.

36 Barbara H. Rosenwein, *Emotional Communities in the Early Middle Ages* (Ithaca, NY: Cornell University Press, 2006); Barbara H. Rosenwein, 'Problems and Methods in the History of Emotions', *Passions in Context: Journal of the History and Philosophy of Emotions* 1:1 (2010), 1–24. See also Andrew Lynch, 'Emotional Community', in Susan Broomhall (ed.), *Early Modern Emotions: An Introduction* (London/New York: Routledge, 2017), pp. 3–7.

37 William M. Reddy, 'Against Constructionism: The Historical Ethnography of Emotions', *Current Anthropology* 38 (1997), 327–51; see also William M. Reddy, *The Navigation of Feeling* (Cambridge: Cambridge University Press, 2001).

38 Rosenwein, *Emotional Communities*, p. 23.

39 Ibid., p. 24.

40 Andreas Reckwitz, 'Affective Spaces: A Praxeological Outlook', *Rethinking History: The Journal of Theory and Practice* 16:2 (2012), 241–58. The sociologist Reckwitz prefers to speak of 'affects' since 'emotion' carries for him a static notion while 'affect', reminiscent of 'to affect' and 'to be affected', has a more dynamic and interactive dimension. Cf. Giovanni Tarantino, 'Mapping Religion (and Emotions) in the Protestant Valleys of Piedmont', *Asdiwal* 9 (2014), 93.

41 Tarantino, 'Mapping Religion (and Emotions)', 93.

42 Benno Gammerl, 'Emotional Styles – Concepts and Challenges', *Rethinking History: The Journal of Theory and Practice* 16:2 (2012), 164.

43 Gauvin Alexander Bailey, *Architecture and Urbanism in the French Atlantic Empire: State, Church, and Society, 1604–1830*, McGill-Queen's French Atlantic Worlds Series, 1 (Toronto: McGill-Queen's University Press, 2018). Bailey is currently working on a second book project entitled 'The Architecture of Empire: France in the Indian Ocean and Southeast Asia, 1664–1954'. See also Gilles-Antoine Langlois, *Des villes pour la Louisiane française, théorie et pratique de l'urbanistique coloniale au XVIIIe siècle* (Paris: L'Harmattan, 2003); Laurent Vidal and Émile D'Orgeix (eds), *Les villes françaises du nouveau monde* (Paris: Somogy, 1999).

44 Gauvin Alexander Bailey, *Colonial Architecture Project*, www.colonialarchitecture project.org, retrieved 13 June 2018.

45 See Bailey, *Architecture and Urbanism*, p. 13.

46 Ibid., p. 4.

47 See, for example, Denis Diderot and Jean Le Rond d'Alembert, 'Ingénieur', in *Encyclopédie, ou Dictionnaire raisonné des sciences, des arts et de métiers*, vol. 8 (Paris: Briasson, David, Le Breton, Durand, 1766), p. 741: 'Nous pensons donc que la perfection de la fortification actuelle est un objet digne de l'attention et de l'application des plus savants ingénieurs.'

48 This function, however, should not be confused with what political scientists term 'social' or 'political engineering', which alludes to the manipulation of a social or political system.

49 Steven J. Jackson, 'Rethinking Repair', in Tarleton Gillespie, Pablo J. Boczkowski, and Kirsten A. Foot (eds), *Media Technologies: Essays on Communication, Materiality, and Society* (Cambridge, MA: The MIT Press, 2014), p. 221–41.

50 Francesca Trivellato, *The Familiarity of Strangers: The Sephardic Diaspora, Livorno, and Cross-Cultural Trade in the Early Modern Period* (New Haven, CT: Yale University Press, 2009), p. 7.

51 See, for example, Jan Vansina, *Oral Tradition as History* (Madison, WI: University of Wisconsin Press, 1988); Adam Jones, *Afrika bis 1850*, Neue Fischer Weltgeschichte, 19 (Frankfurt am Main: Fischer, 2016).

52 Most of the material is found in the Dépôt des fortification des colonies (DFC) in the Archives nationales d'Outre Mer (ANOM) in Aix-en-Provence; there is also a large collection of maps and plans in the Département cartes et plans of the Bibliothèque national (BNF) in Paris. For the French Antilles private collections of maps, for example, on Martinique, see the collection of Jeff Bodington (www.patrimoines-martinique.org/?id=43, retrieved 23 August 2017).

53 The administrative correspondence and collection of memoirs extends over many centuries and includes the series of the department of navy and colonies, especially the series of COL and MAR in the National Archives in Aix (ANOM), but also the central archives in Paris (CARAN).

54 For the Caribbean case, see Alistair J. Bright, '"Removed from the Face of the Island": Late Pre-colonial and Early Colonial Amerindian Society in the Lesser Antilles', in Corinne Lisette Hofman and Anne van Duijvenbode (eds), *Communities in Contact: Essays in Archaeology, Ethnohistory & Ethnography of the Amerindian Circum-Caribbean* (Leiden: Sidestone Press, 2011), pp. 307–25.

55 See, for example, William Moss, 'The Archaeology of a North American and the Early Modern Period in Québec', *Post-Medieval Archaeology* 43:1, Special Issue: The Recent Archaeology of the Early Modern Period in Québec City (2009), 1–12.

56 The œuvre by Ibrahima Thiaw is particurarly interesting for the social and political dimension of spatial construction of a society affected by a colonial encounter: Ibrahima Thiaw, 'An Archaeological Investigation of Long-Term Culture Change in the Lower Falemme (Upper Senegal Region) A.D. 500–1900' (PhD Dissertation, Rice University, 1999); Ibrahima Thiaw, 'Every House Has a Story: The Archaeology of Gorée Island, Sénégal', in Livio Sansone, E. Soumonni, and Boubacar Barry Africa (eds), *Brazil and the Construction of Trans-Atlantic Black Identities* (Trenton, NJ: Africa World Press, 2008), pp. 45–62; Ibrahima Thiaw, 'L'espace entre les mots et les choses: mémoire historique et culture matérielle à Gorée (Sénégal)', in Ibrahima Thiaw (ed.), *Espace, culture matérielle et identités en Sénégambie* (Dakar: Codesria, 2010), pp. 18–41. See also Christopher R. DeCorse, *An Archaeology of Elmina: Africans and Europeans on the Gold Coast, 1400–1900* (Washington, DC: Smithsonian Institution Press, 2001).

57 Colbert to de Baas, 8 February 1674, in Médéric-Louis-Elle Moreau de Saint-Méry, *Loix et constitutions des colonies françoises de l'Amérique sous le vent*, 6 vols (Paris: Moutard et al., 1784–90), vol. 1, pp. 784–5.

[19]

1

Colonial enclosure: Fortification and castles on the Lesser Antilles

Islands fortified

Is it possible to build an empire on an island? From a new materialist perspective on the formation of imperial spaces this may be so.[1] After the arrival of French entrepreneurs and the establishment of settlements on some Caribbean islands, architectural structures transformed landscapes, and spatial representations like maps and plans contributed to the construction of a permanent imperial space. The process of spatial construction will be followed more closely for the cases of Martinique and Guadeloupe, two islands that were colonized by French engineers and cartographers before colonization proper. The strategy proposed by these protagonists of colonial enclosure was based on spatial practices that largely rested on ideas of transposing European concepts of fortification and defensive buildings to the islands.

Based on ideas in postmodern geography this chapter shows that it would be wrong to presume a binary model of spatial construction that perpetuates oppositions of inclusion and exclusion, the 'we' and the 'other'.[2] Instead it shows that the French strategy of imposing their material culture did not result in a space that resembled the centre–periphery dichotomy. Rather this space has to be understood in its own right. Although being confined to the outside it was rather open to the interior, mixing European spatial and material elements with indigenous ones in this sense may be described as a hybrid space within a colonial setting.

In 1669, Jean-Charles de Baas-Castelmore, first governor-general of the island of Martinique, wrote a memorandum to Jean-Baptiste Colbert, head of the royal administration for commercial and colonial affairs.[3] In this correspondence he reports on the plans of François Blondel, a military officer and a royal engineer, who was sent to the Antilles three years earlier. Blondel had visited Martinique and Grenada, along with the adjacent Leeward Islands of Guadeloupe and Saint-Christophe, colonized by the French over the course of the last three decades, and also Tortuga, the small island off the northern coast of Hispaniola. He conducted geographical surveys of these islands in order to assess their potential for the installation of a comprehensive system of fortifications. The string of fortresses Blondel planned to build for Martinique, Guadeloupe, and Saint-Christophe was, in fact, so extensive that de Baas compared this enterprise to the chain of strongholds at the French frontier with Flanders:

> He wants [to build] great fortifications equipped to a high standard with large, deep wells, half-moon fortifications and horn-shaped bastions at precise and regular intervals to the extent allowed by the situation and the rules of engineering, as if he wanted to construct the citadels on the borders of Flanders, and each had to survive a siege against a large army.[4]

The governor was sceptical about the feasibility of such a large building project, which involved hundreds, perhaps thousands, of workers and great quantities of tools and materials, such as stone, chalk, and wood, as well as provisions, which would all cost the government considerable sums of money to be paid over many years to come. De Baas appreciated Blondel's expertise and the job he did of drawing up maps of the islands. His descriptions were very exact, particularly those depicting natural formations that would make good sites for ports, roads, and fortifications. He suspected, however, that if one were to build all the fortresses the engineer had in mind for these places the costs would not only be excessive, but the plan would also fail to take advantage of the many natural defences of the island. In the case of an enemy attack, the inhabitants were able to retreat to the natural protection of certain places, avoiding a direct confrontation or a formal siege, and to wait until the invasion ended due to the lack of supplies needed for a more sustained attack. What Blondel intended instead, de Baas seems to imply, was to implement the art of fortification on these islands to perfection, which would be, as he writes, a project doomed to fail since it lacks the necessary time and funds.[5]

The memorandum brings up the question as to what the French were actually going to do to secure their colonial presence in the Caribbean. Blondel's plan goes further than just implementing a military defence for the islands. For him, Blondel writes in his own statement on Guadeloupe, the terrain of the island provides ideal conditions for a rich agriculture, since the mountains are high enough to make it rain sufficiently for the agricultural labour of the inhabitants to be rewarded abundantly.[6] At this time the western part of the island of Guadeloupe, called *Basseterre* by the French, was divided between two proprietors, who bought it after the *Compagnie des Iles de l'Amérique* went bankrupt in 1646.[7] One was Charles Houël, seigneur du Petit Pré, who was also governor of Guadeloupe from 1643; the other buyer was his brother-in-law, Jean de Buisseret, who also held the office of royal councillor.[8] Both had erected castles to underline their seigniorial supremacy of the island. The fortresses, however, were not situated to overlook the two halves of the islands, but rather close together on the southwestern coast. Houël's fortress, Fort Saint-Charles, was located by the Galion River, while Boisseret's Fort de la Madeleine stood on the banks of the Baillif River.[9] In 1664, the properties were sold to the newly founded West India Company, a few years before Blondel was sent there to examine the state of the two fortresses.

One can gain several insights from Blondel's description of the buildings. He first paints an unfavourable picture of the castle at the Galion, a fortress in the form of an octagonal star, but with imperfections, rather small and with very old walls liable to crumble if hit by enemy cannon fire. Houël's castle was, in fact, 'quite an appalling tour carée', Blondel writes, 'so small, so badly maintained, and, to say the truth, so ridiculous, that this building appears to be nothing more than a heavy mass of stones stacked together with neither intent nor order'.[10] In this state Fort Saint-Charles would serve as nothing more than a prison. Blondel estimates that the substance of the existing structure would not justify modernizing the fortress by employing thousands of men, only to stumble from one problem to another.[11]

This view contrasted directly with that of Père du Tertre, the author of the famous *Histoire générale des isles de S. Christophe, de la Guadeloupe, de la Martinique et autres dans l'Amérique*, who held the defensive works of Charles Houël in somewhat higher regard.[12] The structure he had built after he bought the island was 'a structure totally new with four façades and four levels; each level had four connected chambers, the walls were built of very nice stones [...]; a battery of six cannons commanded the harbour and could thus defend the castle against approaching ships'.[13] The fortress was to underline Houël's pretension of being the absolute master ('le Seigneur absolu') of Guadeloupe.[14] As such the building represented more a symbolic conception of a medieval feudal land order that Houël had to maintain against interest from the investors in the ancient company, other French inhabitants, foreign refugees, especially those Jewish and protestant migrants who fled from Recife in formerly Dutch Brazil after 1654,[15] and, perhaps most pressingly, against the threat posed by attacks of the indigenous population and African maroons in the same year.[16]

Especially mindful of English battleship fleets, Houël began fortifying locations where invaders could make landfall. He ordered the felling of trees along the shoreline in order to block the roads into the interior and to use the trunks for defences. He also mounted cannons at elevated positions and erected – probably only provisionally – batteries that could protect, du Tertre tells us, the whole of the coastline.[17] The defensive works, however, were not tested, since the approaching English Republican fleet under Admiral Sir William Penn failed to lay siege to Guadeloupe and instead continued its mission to eventually capture Jamaica for the Commonwealth.

The Fort de la Madeleine, on the other hand, was much better suited to serve as a basis for Blondel's plan to systematically fortify the island of Guadeloupe. It was in much better shape and had, in the engineer's opinion, the potential to be transformed from 'country house' into a considerable fortress that an enemy wanting to make himself master of the island would not dare to lay siege to.[18] Blondel proposed, Kissoun notes, to turn the old building into an 'impenetrable fortress in the image of those in France';[19] if it had been laid out by only a slightly more intelligent person and 'according to the rules of good fortification' there would have been nothing wanting.[20]

But the proposals were not followed after Blondel returned to France, where he continued his remarkable career as a royal architect in Paris. There he built, among other projects, the Porte Saint-Denis, a monumental landmark that still stands today.[21] The verdict of governor-general de Baas that Blondel's projects were altogether too expensive indicated the manner in which the fortification of Guadeloupe and the other islands would continue to be pursued. In 1690, Charles de La Roche-Borbon de Blénac, the new governor-general of the Antilles, ordered Pierre Hinselin, governor of Guadeloupe, to raze the Château de la Madeleine since the assembly decided that there were insufficient troops to equip both fortresses.[22]

Blondel's brief sojourn in the Antilles has thus not left its mark in the architectural history of the islands. He did, however, express an idea within his plans and writings that proved influential to the whole understanding of colonial space as a territorial entity. Historians have noted that Blondel's plans to secure Guadeloupe applied not only the technical 'rules of good fortification', but also the same principles that fortification engineers were following on the northwestern borders of France.[23] But one can go a step further in interpreting Blondel's plans as an attempt to colonize through the creation of territorial space. The naturally given geographical borders between land and sea were not sufficient to transform the island into French territory. The frontiers of colonial space had to be, just as in France, constructed as cultural,

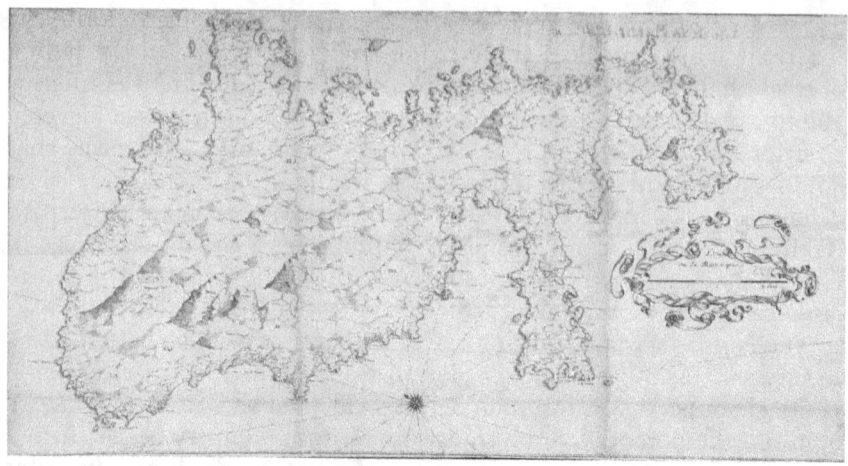

1   François Blondel: L'isle de la Martinique, 1670: a bird's-eye view of the island from the East that shows the steep contours of the shoreline to highlight the places suitable for fortification.

linguistic, and physical borders as a result, not as the origin, of political interaction.[24] David Bitterling has recognized the role the idea of absolute space had played in the thought of Sébastien Le Prêtre de Vauban, the famous engineer of fortification in constructing France's space as territory.[25] In fact, Vauban's idea of France as *pré carré*, the enclosure of its territory by an 'iron belt' of fortresses, as well as its application of a manorial economy,[26] can equally be superimposed onto the colonial setting of Blondel's Guadeloupe.

A look at the map Blondel sketched for the island of Martinique gives an idea of how the engineer carefully highlighted the buildings that bordered the ragged coastline (see Figure 1). The plots of land on the island that were granted by the King to individual settlers had an elongated rectangular shape, according to Alexandre Oexmelin, mostly 400 by 60 *pas géométriques*, with the residence built near the sea.[27] Blondel depicted the houses of these properties, but not the enclosed land that surrounded them. Instead, he noted the name of the owner next to each house symbol, all of which are shown evenly spaced one from the other. Sometimes properties extended over a whole bay, which is why Blondel gives them the name of the habitation. Interestingly, however, he does not choose a different symbol for the dwellings of the indigenous people. He rather merged the French habitations with the settlements of the Caribs. Sometimes indigenous and French houses are drawn closely together forming a collection in the manner of a village – for example, at the Pilote River near the Pointe de Cul de Sac

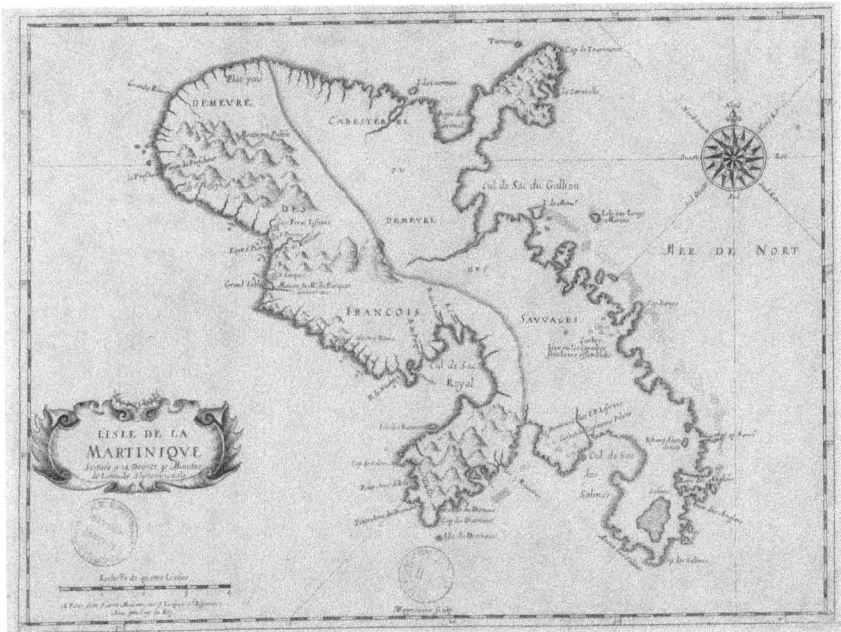

2   Nicolas Sanson: L'isle de la Martinique (Paris: Pierre Mariette, [ca. 1650])
(BNF Paris, Département Cartes et plans, CPL GE DD-2987 (9102)).

(today's Pointe Borgnesse) and present-day Anse Figuier. There, one can see two houses with the description 'Caraibes' and one denoted as 'La Pere'.

There is no division of territories on the island on Blondel's map as Nicolas Sanson has visualized it on a printed map of Martinique from around 1650 that draws a line through the interior of the island (see Figure 2).[28] Sanson located a Carib village at the Pilote, where one can find mentioned a 'Carbet du Capitaine Pilote' as well as the 'demeure des P.P. Jesuites'; there is another village at the Cap Louis, near present-day Le Vauclin, a 'Carbet', that is, an indigenous hut or a small settlement, 'the place where the Caribs have their assembly'.[29]

A later map with the same line of division by Nicholas Visscher from around 1680 shows even more Carib settlements, for example, a 'Case de Caerman', a Carib chief, situated near present-day Sainte-Marie on the north-east coast (see Figure 3).

In 1704, a map by Nicholas de Fer registers yet another assembly of houses glossed as 'Carbet de Sauvages Macabou', a village under a chief called Simon by the French, but who was probably called Macabou among the Caribs (see Figure 4).[30]

[25]

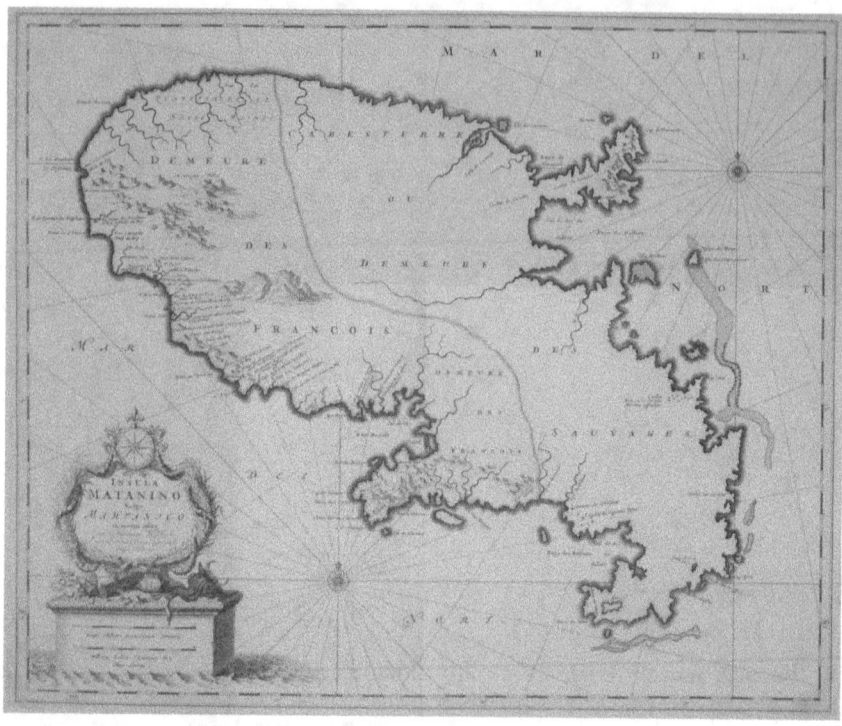

3 Nicolas Visscher II: Insula Matanio vulgo Martinco in lucem edita per Nicolaum Visscher (Amsterdam: Nicolaus Visscher, [ca. 1680]).

The French names given to indigenous chiefs and thus settlements appear regularly on Blondel's map: the bays de Louis, de François, de Simon, de Robert – they all sound like French proprietors, but were, in fact, indigenous places, stripped not only of their original name, but also of their territorial independence. It is telling, therefore, that Blondel does not note any indigenous name on his map, while de Fer, for example, denotes a cape that Blondel leaves unnamed as Pointe du Macabou.

Blondel's cavalier technique of depicting the island from a bird-eye's perspective underlines the effect of the island as a homogeneous entity. Its precipitous coastlines figure as ramparts that serve as natural defences and the mountains in the interior convey a sense of inaccessibility. Perhaps the most prominent cliff was the rocky peninsula of Fort Royal that separated the small bay of Vasseur and the Carenage in the Cul-de-Sac-Royal (see Figure 5). Even though the fortress was not yet expanded to today's state, Blondel has depicted its site quite dramatically to give the impression of a mighty stronghold, which, in fact, it was not at

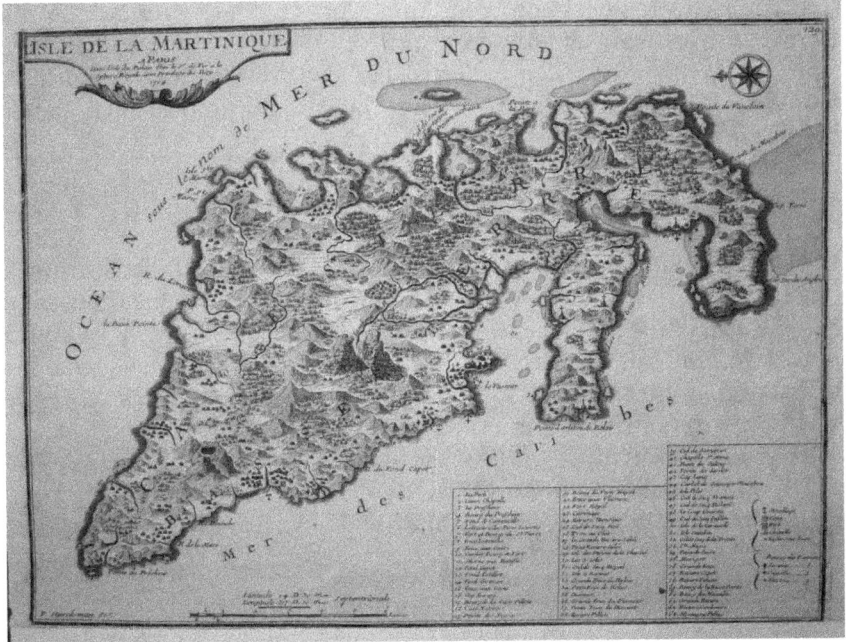

4 Nicholas de Fer: Isle de la Martinique (Paris: de Fer, 1704).

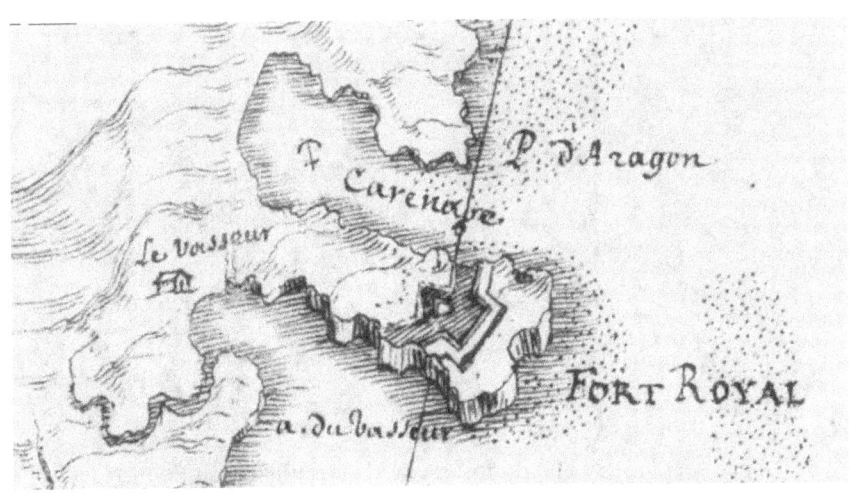

5 Blondel: L'isle de la Martinique, Detail of Fort Royal.

that time. But his design to construct a new fortress overlooking the Carenage was again dismissed as not only being very costly for the King, 'more than fifty thousand écus', but also useless, since the fortress would be misplaced, guarding neither the harbour nor the entry to the bay from above.[31]

Walls of power

During the seigniorial period the great proprietors of the French Antilles pursued a building strategy of representation.[32] The manor houses of the first half of the seventeenth century, therefore, did not serve defensive purposes, but rather to convey a sense of a symbolic power over the island and its inhabitants. Some examples of such buildings may give an impression of the power practices involved with these castles.

I have already mentioned the Château de la Madeleine on Guadeloupe, of which Blondel sketched a view that showed the surrounding terrain and the close-by village at the Baillif River (see Figure 6). Blondel depicted

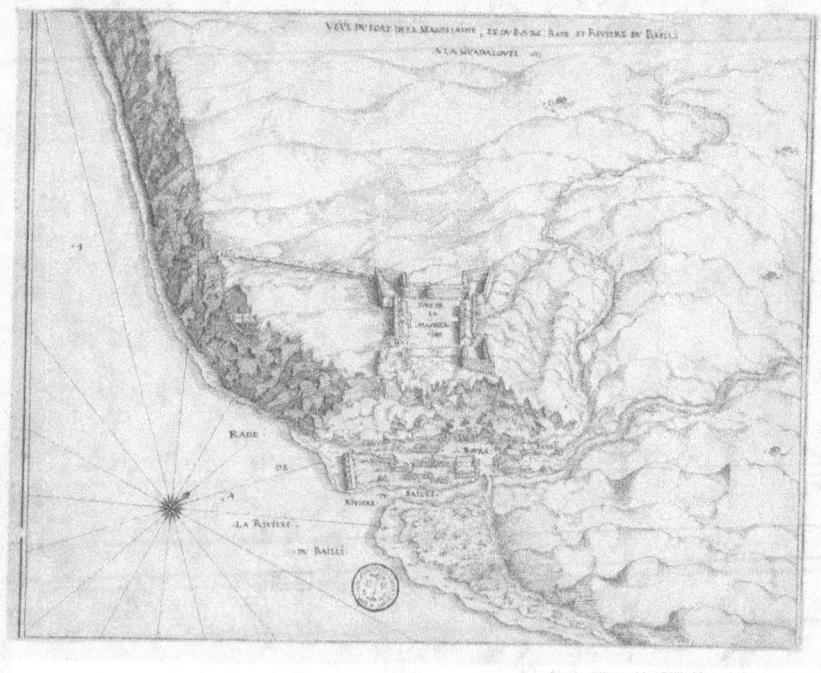

6   François Blondel: Veüe du fort de la Magdelaine, et du bourg, rade et rivière du Bailli, 1667 (BNF Paris, Département Cartes et plans, GE SH 18 PF 155 DIV 4 P 2 D).

the fortress as particularly unspectacular, both from an engineer's perspective and from a representational viewpoint.

The stone walls were narrow, the general shape in the ideal form of a square of which one bastion was missing, and inside there were only four simple housing structures lacking any representational value. But that was probably not what its owner Jean de Buisseret had in mind to convey to the observer. The Pere du Tertre, on the other hand, included a picture of Fort de la Madeleine in the second edition of his *Histoire général des Antilles*, which showed the fortress in its complete geometrical form as well as with a representative portal and residence inside, the latter having an elaborate façade and a long balcony (see Figure 7). Charles de Rochefort, mostly at odds with the Dominican, describes its appearance equally favourably:

> The Governor made a castle out of his dwelling that was not far from the village. He had built it solidly with four faces. The corners were armed with jetties and redoubts, the masonry of such a thickness that it could withstand the weight of several cannon shots [...].[33]

The verdicts were equally divergent in the case of the Château de Poincy on the island of Saint-Christophe, known today as Saint-Kitts.[34] It is the second castle depicted in the engraving mentioned in the *Histoire générale* that seems to represent a proper castle in French style surrounded by a geometrically ordered garden. The Château de la Montagne, as it was also known, stands as one of the earliest examples of the baroque architectural style brought to a French colony. After most of the structure turned into rubble due to an earthquake, Jean-Baptiste Labat, the famous early historian of the French Antilles, visited the place. He mentions the elevation of the terrain that the castle was built on, that it was very beautiful, and that it had a commanding view over the adjacent countryside. Labat felt that the castle, according to the picture in the du Tertre book, must have been magnificent, an impression he was able to confirm with his own eyes as he observed the ruins lying among its many terraces, signifying the wealth and good taste of the man who had built such an edifice. There were still some intact grottos, basins, lead pipes, and reservoirs for a fountain that drew its water from a higher mountain half a mile away.[35]

Labat at this time did not regret so much the loss of seigniorial splendour, but rather the lost opportunity to reactivate the hydraulic system for public use. He proposed to build aqueducts or water pipes in order to supply the nearby village, which was in great need of such assistance, being reliant solely on rather inadequate cisterns and wells. He was already envisioning putting the African slaves on the island to

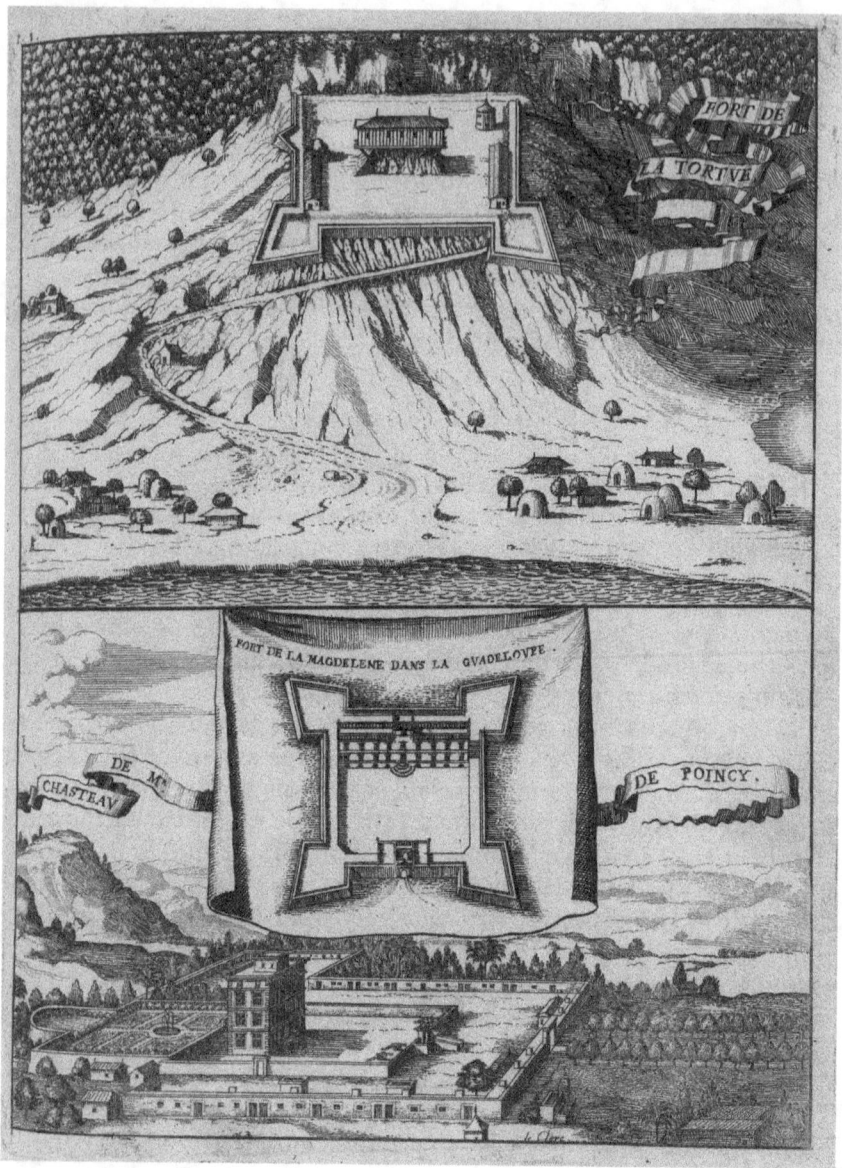

7 Three French castles of the Caribbean typical for the period before 1660:
Fort de la Tortue or de la Roche on Tortuga, Fort de la Madeleine on
Guadeloupe, and Chateau de M. de Poincy on Saint-Christophe, in Du
Tertre, *Histoire générale des Antilles*, vol. 2, p. 4.

work on such projects. He also underlined the absolute necessity of their manpower for the well-being of the island's general economy.[36]

In the opinion of Charles de Rochefort the Château de Poincy surpassed all the other castles on the island. He, too, praised the advantageous location, the pleasant appearance of the adjacent houses for officers and settlers, and the avenues bordered with orange and lemon trees. But most of all, Rochefort liked the 'beautiful Palais':

> Its outline is nearly square; with three nicely proportioned floors that follow the rules of an exquisite architecture, employing stone and brick masonry with a nice symmetry. The façade that presents itself first looks to the east, has a large staircase in front of the entrance, leading down on either side, and a nice parapet above; the one with a west-facing aspect is also embellished with a staircase just like the first one, as well as with a beautiful and plentiful spring of rushing water that is received by a large basin and from there carried on to all points via subterranean canals.[37]

Inside, the halls and chambers were well laid out, the floors made in the French fashion of red wood, being solid, polished, with a pleasing smell and originating from the island. The rooftop served as a terrace with the most wonderful views; the windows to the front were placed in a nice order looking out over the length of the avenue and the small valleys beyond where sugar cane and ginger grew. To the back the house faced the mountains, which were at just the right distance to form a view in perfect proportion with the Palais.[38]

The garden itself was perhaps the most remarkable feature for a colonial castle representing French style in the Antilles. Rochefort found it furthermore well tended. For the most part, the garden was planted with herbs and vegetables that had been brought from France. Rare and curious flowers surrounded the fountain that sprinkled water from the mountain source situated higher up. Rochefort continues by imagining how people would have celebrated news of the victories of Louis XIII with fireworks in this setting. Music played on the roof terrace would have resonated around the neighbouring mountains, forming a pleasant echo that would extend over the whole island and well out to sea. All this would represent the style of French courtly life on Saint-Christophe, symbolized by the Fleur-de-Lys emblazoned high up on the front façade as well as on the flags and banners the General would have carried out to meet his enemies.[39]

The highly representative castle was also fortified: redoubts, large cannons, and an arsenal where all sorts of arms and supplies of gunpowder, fuses, and bullets were stored in abundance. Next to the military installations were those for agricultural purposes, that is, the sugar mills, operated by African slaves, 'who rendered their master a profit and a secure revenue on equal terms with the most noble and best seigniories of France'.[40]

The ravishing portrait of Poincy's castle that Rochefort sketches for his reader can be given some credence even though the Pere du Tertre finds the description somewhat exaggerated. It might have been the 'most beautiful house of the island', he writes, but its splendour is not as extravagant as Rochefort would have it. The castle itself does not measure more than 7 or 8 toises, that is, about 14 or 16 m, in height and the surrounding buildings, for example the arsenal, are only built either of brick or, in the case of the chapel, of wood. In the African village nearby, called Ville d'Angole, however, there were some stone and brick buildings that housed several artisans.[41]

The inventory of Poincy gives us an even better picture of a seigniorial household in the Antilles.[42] The Château de la Montagne was well equipped with expensive furniture, mirrors, paintings, tapestries, expensive Dutch fabrics, a gold chain worth a thousand livres, and even two globes and some maps, which were located in a second-floor room facing south (see Figure 8).[43] Most interestingly, however, is the library with about 150 books containing, among other things, works on hydrography, geography, and some recent volumes on fortification.[44]

Poincy represented, therefore, not only the seigniorial type of the first period of colonialization in the Antilles, but was in fact already pursuing the further aim of securing the French presence in the region against other European nations.[45] On the one hand, he introduced a French style to the island that was intended to appeal to all the senses of its inhabitants, and was not only directed towards the neighbouring English adversaries, but was also designed to establish a certain emotional community amongst French, African, and indigenous groups.

On the other hand, the spatial conception of the Château de la Montagne already displayed some features in common with the new absolute order of space introduced later on by engineers like François Blondel. Perhaps most indicative of this is the wall, with its representative portal, that surrounds the main building. Neither Rochefort nor du Tertre nor Labat included this architectural feature in their descriptions. It may have been such a common sight for the authors that they have just depicted it visually. But the walls signified the territorial enclosure, hinted at further above, that was characteristic of the plantation system on the French Antilles. The descriptions of the Château and its surrounding agricultural landscape can thus be seen as an ideal model for the manorial economy of each habitation (see Figure 9).[46]

Tortuga (île de la Tortue) was the fourth island French adventurers – the famous buccaneers and filibusters – took possession of in the

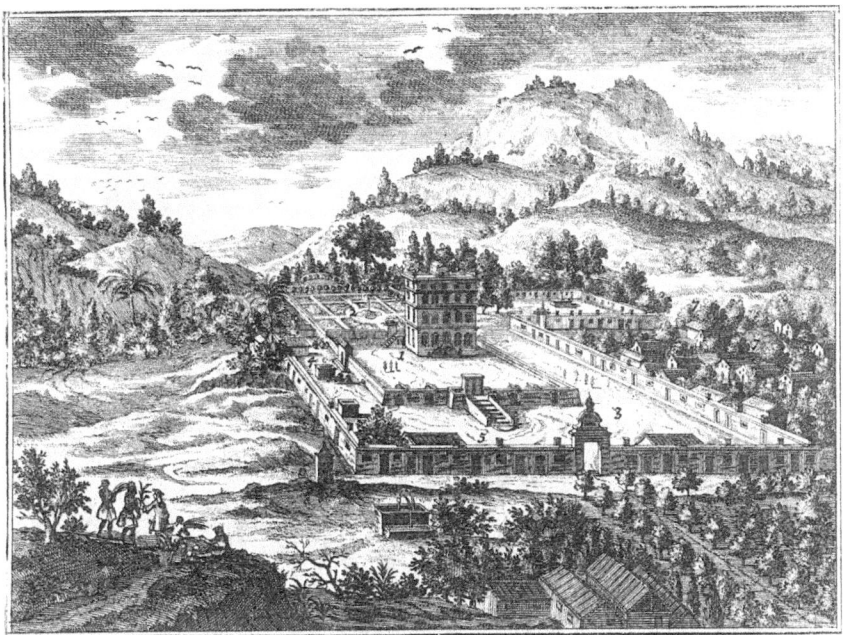

8 The Château de la Montagne commissioned by the governor-general of the French Antilles Philippe de Longvilliers de Poincy around 1640. The engraving is taken from a later edition of Charles de Rochefort's *Histoire naturelle et morale des îles Antilles de l'Amérique* (Rotterdam: Arnould Leers, 1681), p. 53.

Caribbean. In 1640, the engineer Jean Le Vasseur was sent by Poincy together with forty or fifty of his Huguenot compatriots and about the same number of buccaneers from nearby Hispaniola to retake the island from the English and Spanish.[47] After the French had driven out the English governor and his entourage, Le Vasseur, now 'absolute master of this island', commenced building a fortress on a rock: Fort de la Roche, overlooking the harbour and 'naturally fortified'.[48] On top of it he built a stone house from the same material as the rock. It served him as an abode – the 'dovecote' – and the only way to access it was via an iron ladder.

The Fort de la Tortue, the third castle that is depicted on the engraving in du Tertre's *Histoire*, is perhaps the least impressive of this small visual compilation of French representative buildings in the Caribbean. In his description of the island, however, du Tertre estimates that the fortress might have been the most powerful in all the French islands.[49] This claim to power obviously was problematic for Poincy's own

[33]

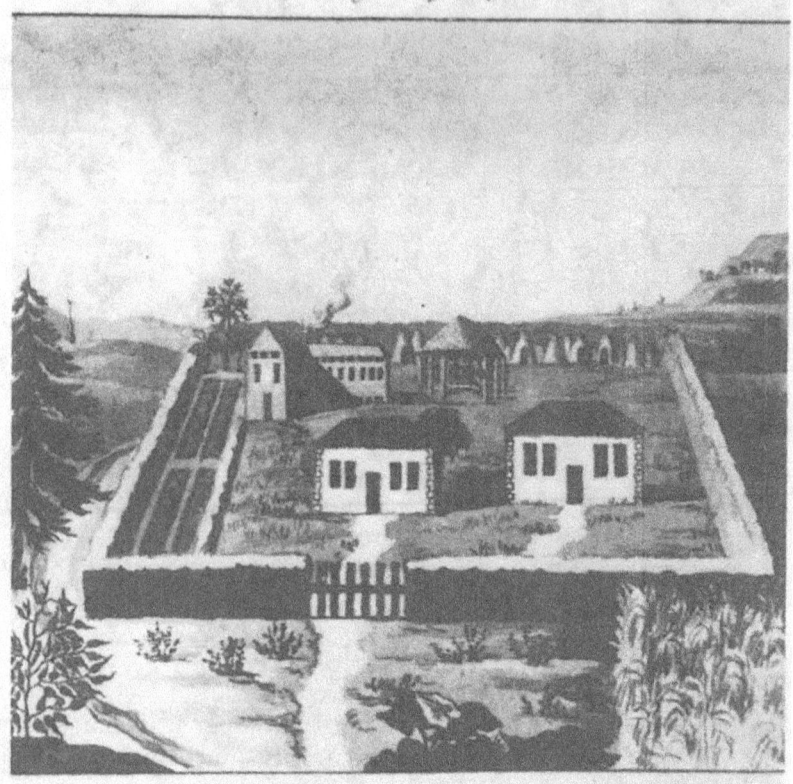

9 Nicolas Halma: Vue d'une habitation situé dans le canton de la parroise de Trois-Islets, près de Fort Royal, à pres de distance de la mer, 1805. This image of a habitation near the birthplace of the French empress Josephine in the town of Trois-Ilet on Martinique clearly depicts the walled enclosure of the property.

pretensions as the highest-ranking governor of the French Antilles. Le Vasseur had promised him, according to the Dominican, to establish a habitation on Tortuga. Poincy, who was well informed about the proceedings of Le Vasseur's efforts to fortify the island, sent a delegation from Saint-Christophe in order to remind him of his promise. The engineer, however, deflected this advance and avoided direct confrontation with nice talk and great civilities.[50] Once he had repulsed a Spanish counter-attack, Le Vasseur began to rule the society of buccaneers in an absolute fashion that caused many grievances. Du Tertre himself was particularly critical of his grasp on power, since the governor not only raised taxes, but also persecuted Catholics on the island, burned one of the chapels, 'that they built with their own hands', expelled two priests, and did not even spare his own Calvinist minister Charles de Rochefort, whom he dispensed from holding service. He imposed harsh punishments on the inhabitants, hanging them up in an iron cage he called 'hell' and imprisoning them in his fortress, for which he coined the name 'purgatory'.[51]

Even though Poincy tried his best to hold Le Vasseur accountable to his government in Saint-Christophe, the latter continued to mock his superior by withholding large prizes of Spanish silver for himself, since Catholics like Poincy, he sarcastically remarks, were so spiritual that they obviously had no need for precious metal. Le Vasseur, however, ruling 'more as a king than a governor', fell victim to a conspiracy. Two of his captains – young men whom he had adopted, before sexually abusing one of their mistresses several times – resolved to assassinate the tyrant. In a scene du Tertre dramatically compares to the assassination of Julius Caesar by Brutus and his conspirators, the patricides attacked Le Vasseur as he descended from his rock, in the company of a group of men, en route to a warehouse, shot him, and after a struggle killed him with their daggers. After the death of the tyrant, the captains, who had already won the loyalty of the inhabitants by promising to let them live in liberty ('de laisser vivre dans toute sorte de liberté'), took possession of the fortress and of all Le Vasseur's property.[52]

As well as standing for the feudal supremacy of a certain land proprietor, a fortress was also able to symbolize the political dominance that a ruler could exercise over his subjects. The case of Le Vasseur and Fort de la Roche might be a particularly dramatic episode of the potential power inherent in buildings, but in the relatively loosely organized and rather unhierachical society of buccaneers, a group of individuals could be subdued by another individual through knowledge of engineering and through expertise in fortification. The rather brutal practice of Le Vasseur's government could, therefore, serve as an example of the use of the psychological advantage of a large and monumental building in order to establish the formal rule of a colonial and state-controlled administration. The independence the governor of Tortuga enjoyed in wielding his power on the island was in part due to the impotence of Poincy, who was not able to extend his government beyond his island of Saint-Christophe.

It was, therefore, of primary concern to the official government on the islands that they extend their control by means of engineering, not only to combat threats from the outside and other European nations, but also to maintain control over a diverse and pluralistic population within the French Antilles. The next appointed governor of Tortuga, Timoléon Hotman de Fontenay, undertook such building projects as the construction of a new chapel to be used to celebrate the Catholic mass as well as the addition of two new bastions to the fortress.[53] The Fort de la Roche was once more tested by a Spanish attack beginning 10 January 1654. Using slaves, the Spanish succeeded in installing a wooden battery with eight or ten cannons on the mountain above the fortress. As a result of the ensuing bombardment, the fort sustained some damage and even the governor's house at the top was hit. But the fortress held against the assault, although the French were eventually deposed from the island by treason in their own ranks.[54]

The settlement of Martinique, started by 'illustrious founder of the French colonies' Pierre Belain d'Esnambuc in July 1635, was the site for the construction of Fort Saint-Pierre and Fort Royal.[55] At first, the fortresses were very simple structures with wooden palisades, but were developed further under Jacques Dyle du Parquet, nephew of d'Esnambuc and new governor of Martinique. There was, as the Chevalier de Poincy reported from Saint-Christophe to the president of the Compagnie des îles de l'Amérique, François Fouquet, only one available carpenter for the refitting of the palisades of Fort Royal and other building projects.[56] Parquet, therefore, asked Fouquet to send artisans, tools, and materials in order 'to build a town':

> Concerning the building of a town, which you have asked me to begin, it is necessary that you send me a number of masons, brick makers, stone dressers (tailleurs de pierre), lime burners, carpenters, cabinetmakers, locksmiths, edge tool makers, nail makers, tilers, and other workers, equipped with suitable tools and other things necessary for the construction of said town, which are not available in this land.[57]

Until the arrival of this cargo, Parquet argues, it will be necessary to continue building 'à la mode du pays', as in the case, for example, of

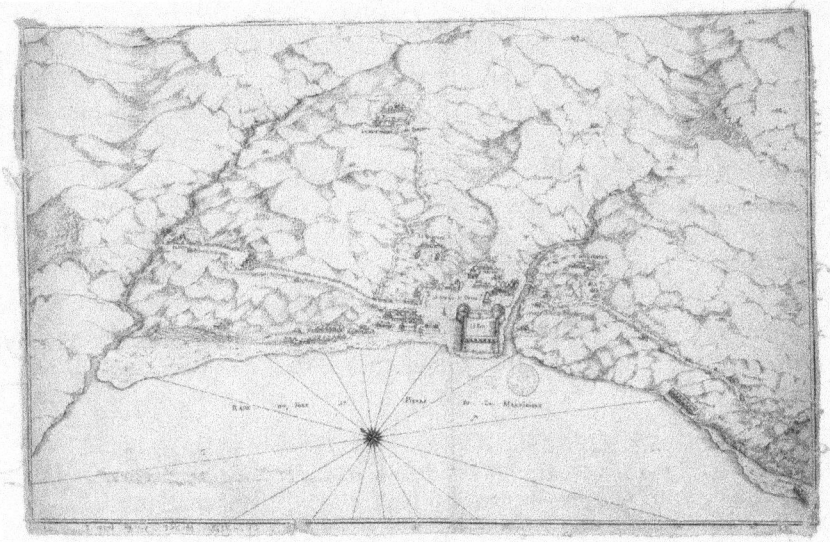

10 François Blondel: Rade du Fort Saint-Pierre de la Martinique, Ms., 1667 (BNF Paris, Département Cartes et plans, GE SH 18/PF 156 DIV 6 P 1 D).

the storehouse of the clerics, which has been covered using only leaves.[58] For himself, however, Parquet had erected, 'at excessive expense', not only a large and beautiful house of stone overlooking Saint-Pierre, but also a second brick building as his seat of government further south at the mouth of the Carbet River (*Quartier du Carbet*).[59]

François Blondel had drawn these places too (see Figure 10). While one can see that Fort Saint-Pierre itself is, indeed, of 'good masonry', as du Tertre tells us, there is a battery for seven (and not nine or ten) cannons at the front of the fortress.[60] The fortress and the village of Saint-Pierre on Blondel's tableau is not a particularly handsome sight. Perhaps the engineer did not care much about the aesthetic dimension Charles de Rochefort emphasizes in his *Histoire*: the Jesuit estate that Blondel depicted outside the town and next to a river together with a watermill is 'a structure that pleases the eye' built of solid stones and bricks, the 'avenues are made quite beautiful', and the town is surrounded by fine gardens and orchards, many plants, herbs, flowers, and fruits brought here from France; there was even a vineyard that produced enough grapes to make a fine wine.[61]

From the Bourg de Saint-Pierre with its central square, where the governor held his council, a path led up to the Montagne de Parquet.

Blondel drew the courtyard enclosed by two long buildings and a wall to the front with a small portal. The mansion itself, compared to other buildings in the drawing, looks larger and more splendid. One can even discern the geometric order of a garden behind the house. This identifies the 'Montagne', as it was called, as the representative residence, similar to the Chateau de Poincy on Saint-Christophe.[62]

The buildings of the Quartier du Carbet, on the other hand, as seen on the map of Martinique drawn by Blondel, are represented in much less detail.[63] The church on the islands between the two branches of the Carbet River can be identified as the former private residence where Parquet lived for seventeen years before he assigned the brick building to the Jesuits.[64] Here, too, Rochefort ascribed to the estate many splendid attributes that du Tertre has denied in his account:

> Monsieur le Gouverneur has honoured this pleasant Quartier [du Carbet] for quite a long time with his residence; he built his house out of bricks situated not far from the bay, near the Place d'armes, in a beautiful valley with quite a large river that came from the mountains. [...] The house itself is surrounded by several beautiful gardens that are bordered by fruit trees and embellished with all the rarities and curiosities of the land. Monsieur le Gouverneur left this dwelling about two years ago, due to his not feeling well in the district in which it is situated, and gave it as a gift to the Jesuits, along with several other nice habitations in its dependency and a great number of African slaves who cultivate the land.[65]

Du Tertre, as sceptical of this description as he is and generally full of doubt regarding the account by Rochefort on many other occasions, found that the gardens 'bordered by fruit trees and embellished with all the rareties and curiosities of the land' were greatly exaggerated and that there was nothing more than tobacco and manioc.[66]

But Rochefort tells us more than du Tertre about the indigenous past of the place that Parquet had first chosen for his habitation:

> The Quartier du Carbet, which has retained its name from the Caribs. These people once had one of their larger villages at this location and a fine cottage that they called Le Carbet, a name that is still present at those places where they hold their assemblies.[67]

The first governor of Martinique had, therefore, intentionally chosen a building site for the first French representative building that was close to, or even within, an indigenous settlement. The purpose might have been similar to that of other building projects designed to direct the material symbolism of power towards the population within the island. But here, we must assume, Parquet also had no choice. In order to make up for his complete lack of resources and workers he had to rely on indigenous infrastructure. Carib and French resources of expertise and materials were more coincidental than it appears from the written sources. In fact, we will see later on that this interaction between interior and exterior agents involved in large building projects laid the foundations for a common and shared identity on the island.

In the years to come, Fort de Saint-Pierre was not to be integrated into a more modern fortification system. There were some developments under Jean de Clodoré, who governed the island under the authority of the Compagnie des Indes occidentales from 1664; also some redoubts were constructed along the coast to guard the bay.[68] But the fortress was, as the governor-general de Baas wrote in 1669, misplaced, and whoever had built it had no knowledge of the geographical situation. It would serve well only as a prison; otherwise it was useless.[69] Instead, a different location on the island in the Cul-de-Sac-Royal seemed much more appropriate for fortification, which de Baas deemed very necessary, since no one, he argued, could call himself master of Martinique and its disobedient and undisciplined inhabitants until these labours were finished.[70]

Even if Blondel's ambitious plans for Fort Royal were not followed through, the simple structure constructed during the 1660s was to become the most important part of the island's defences. In 1694, when Père Labat visited the town of Fort Royal, the fortress was already an imposing sight and had endured a major siege by a Dutch fleet under Admiral de Ruyter in 1674. Labat almost became involved in the planning and construction of a new fortress when the Comte de Blénac, then governor-general of the islands, tried to convince him to stay as chief engineer in Fort Royal. But he declined the offer and left the project in the hands of the naval engineer Jean de Giou de Caylus, in whom he had every faith as an expert, and who received his professional training in Europe.[71] On his guided tour through the fortress Labat realized the grave problems Caylus and Blénac were confronted with. The defensive works were not only damaged, but also incorrectly planned by the former engineer. The natural defence of the rock, in fact, was not sufficient since it extended only over half the peninsula. An earthen slope was protected by a mere double row of wooden palisades defended by two batteries of cannon. Labat argues that lucky circumstance had prevented a successful attack by Ruyter rather than the quality of the defence, which he deemed amateurishly executed.

But Labat's critique of the engineer Payen, who had formerly been in charge of Fort Royal, falls short. It was, in fact, also the case that financial resources for the fortification of the town near the Cul-de-Sac, as well as the availability of material and labour, were just as low as in the vicinity of Saint-Pierre for most of the latter half of the seventeenth century. Labat himself informs his reader of the dire conditions a few decades before his sojourn. There had been just a few cottages built of reed in the swampy area where the later town of Fort Royal was going to be. The building of the fortress of Fort Royal was an endeavour that took many decades and several generations of engineers and workers. The slow progress of this project with all its delays, setbacks, and problems will be described in further detail in Chapter 3.

Over the course of the eighteenth century, however, the French Antilles became more and more fortified. The military struggle with England made it necessary to protect the islands with strong fortresses, batteries, redoubts, and other defensive works. These building projects were pursued despite the large costs and the logistical difficulties in providing the construction sites with enough workers, tools, and materials. After the Seven Years' War the efforts intensified once more, since the weaknesses of the older fortifications had been revealed during this conflict. There were several strong and sophisticated fortresses on the islands that remained in French possession at the turn of the century.

Besides Fort Royal, later called Fort Saint-Louis, and Fort Saint-Pierre there was a small fortress on the north coast protecting the bay of La Trinité (built 1713), and near to this town a large house with an extensive complex of buildings, the Château Dubuc (built 1721). After 1763, Fort Bourbon was built (1764–1772) on the Morne Garnier above Fort Royal. Fort Louis (built 1692), Fort Saint-Charles (formerly Fort Charles and later Fort Delgrès, built 1720–59), and Fort Fleur d'Épée (built 1750–63) appeared on Guadeloupe. There was another Fort Royal in Grenada (later Fort Saint-George, built 1666; rebuilt 1705–10), whilst Fort Saint-Michel guarded Cayenne (today's Fort Cépérou in French Guiana, built 1652), and several fortresses stood in Saint-Domingue, today's Haiti (Fort Dauphin, formerly Fort Bayaja, now Fort Liberté, built 1730–5; Fort Saint-Joseph, built 1755 (see Figure 11); Fort Saint-Frédéric, built 1740; Fort Saint-Charles, built 1740–56; Fort de la Bouque, built 1736; and the Citadelle La Ferrière, built 1820, the foremost example of Haitian engineering in the early period of independence).

The history of these fortresses is generally one of a military expansion of the French presence in this region. Castles originally built to display the power and the fortunes of individual proprietors or governors fell into ruin and were dismantled in order to transform them into modern defence works. This transformation accompanied the demise of the seigniorial period in the French Antilles towards the end of the seventeenth century. Increasingly, fortresses were no longer built by rich feudal lords for representational purposes, but planned, constructed, and maintained at the expense of the royal navy and thus the French state. This trend had consequences for the outward appearance of the

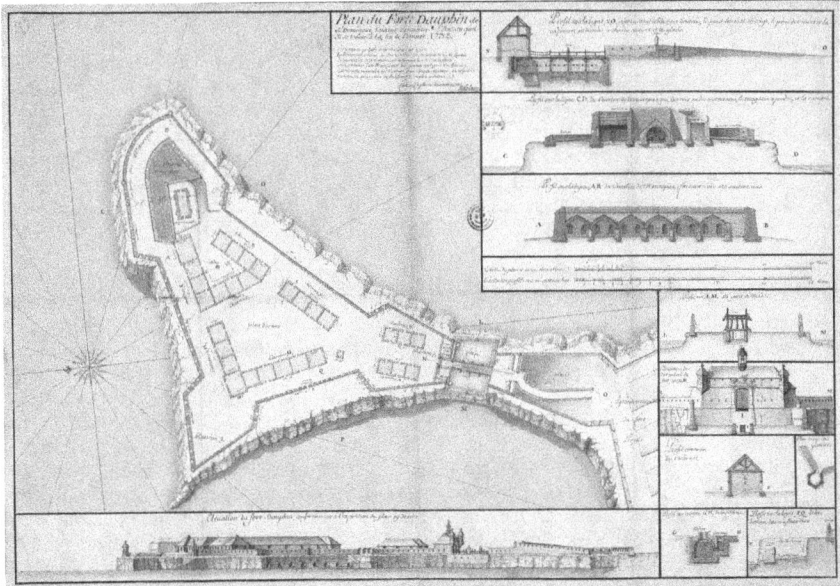

11 Louis Joseph La Lance: Plan du Fort Dauphin de Saint-Domingue
faisant connaître l'état auquel il se trouve à la fin de l'année 1732 (CAOM
15DFC292B). The plan is a good example of the representative quality of
some structures displaying royal symbols and fortification architecture
mixed with French classical style. The main gatehouse is flanked by two
guérites, a small turret with a bell, and ornaments like Fleurs-de-lys and
blazing stone spheres above the gate.

buildings. Geometrical order and tactical efficiency drove the engineers'
plans for the development of the fortresses, whilst ornamental detail,
designed to create a courtly atmosphere like that of the Château de
Poincy on Saint-Christophe, was less of a factor.

The process of fortifying the islands was long and arduous. One
could describe the 'capitals' of the respective French colonies as a
permanent construction site over a period of nearly a century. But
despite the difficulties of establishing a material presence in the Antilles,
the construction practices reveal that the French, Carib, and African
individuals involved in the process managed to resolve many of the
problems. Chapter 3 will take a closer look at the administrative sources
in the colonial state archives, which provide insight into these practices
and paint a more detailed picture of the history of large building projects
in the French Antilles. But for a change of a scene, the next chapter
turns to Pondichéry, the centre of the French empire in the Indian
Ocean.

[41]

Notes

1 Henri Lefebvre, *La production de l'espace* (Paris: Gallimard, 1974); published in English as *The Production of Space*, trans. Donald Nicholson-Smith (Malden, MA/ Oxford: Blackwell, 1991).

2 David Harvey, *Social Justice and the City* (London: Arnold, 1973); Edward W. Soja, *Postmodern Geographies: The Reassertion of Space in Critical Social Theory* (London: Verso, 1989); Michel Foucault, 'Of Other Spaces', *Diacritics* 16 (1986), 22–7; Michel de Certeau, 'Pratique d'espace', in *L'invention du quotidian*, ed. Luce Giard, vol. 1: *Arts de faire* (Paris: Gallimard, 1990), pp. 139–91; Edward W. Soja, *Thirdspace* (Malden, MA: Blackwell, 1996).

3 For general accounts on the French Antilles, see Philip P. Boucher, *France and the American Tropics to 1700: Tropic of Discontent?* (Baltimore, MD: Johns Hopkins University Press, 2008); Paul Butel, *Histoire des Antilles françaises, XVIIe–XXe siècle* (Paris: Perrin, 2007); James Pritchard, *In Search of Empire: The French in the Americas, 1670–1730* (Cambridge: Cambridge University Press, 2004); Jean-Pierre Sainton (ed.), *Histoire et Civilisation de la Caraïbe: Gouadeloupe, Martinique, Petites Antilles. La construction des sociétés antillaises des origines au temps présent: Structures et dynamiques*, Tome 1: *Le temps des Genèses; des origines à 1685* (Paris: Karthala, 2004); Pierre Pluchon, *Histoire des Antilles et de la Guyane* (Toulouse: Privat, 1982); Pluchon, *Histoire de la colonisation française*, vol. 1, pp. 369–431; see the older literature by Nellis M. Crouse, *French Pioneers in the West Indies, 1665–1713* (New York: Columbia University Press, 1940); Cabuzel-Andréa Banbuck, *Histoire politique, économique et sociale de la Martinique sous l'ancien régime (1635–1789)* (Paris: Librairie des sciences politiques et sociales, 1935; repr. Fort-de-France: Société de distribution et de culture, 1972).

4 CAOM COL C8B 1, No. 22: Mémoire sur les fortifications de Martinique, Guadeloupe et de Saint-Christophe, envoyé par M. de Baas, gouverneur général, à la suite de la mission accomplie par l'ingénieur Blondel, 1669, fol. 1r: 'Il veut des grands bastions bien revestus avec des fossé larges et profondes, Des demy lunes et des ouvrages a corne regulier et corects autant que les sçiuations pourront favoriser les regles de l'art, comme s'il vouloit bastir des citadelles sur les frontieres de Flandre, et que chacune deut soustenir un siege contre une grande armée.'

5 Ibid.: 'Ainsy je ne suis pas du sentiment dudit Sr. Blondel, pour faire mettre en perfection les travaux, qu'il a tracé dans les trois isles principales, Il y faudroit employer trop de temps et trop de despence et pour eviter ces deux choses importants, j'escriray ma pensée sur ce qu'il y auroit a faire en chacune des trois isles.'

6 CAOM 08DFC/26, No. 2: Avis sur l'isle de la Guadeloupe et l'estat present de ses fortiffications, par Mr. Blondel, sans date [1666].

7 For the history of this early French colonial trade company, see now Eric Roulet, *La Compagnie des Iles de l'Amérique (1635–1651): Une entreprise colonial au XVIIe siècle* (Rennes: Presses universitaires de Rennes, 2017); for an overview of all the individual and collective proprietors in the Caribbean, see Philip P. Boucher, 'French Proprietary Colonies in the Greater Caribbean, 1620s–1670s', in L. H. Roper and B. Van Ruymbeke (eds), *Constructing Early Modern Empires: Proprietary Ventures in the Atlantic World, 1500–1750* (Leiden/Boston, TX: Brill, 2007), pp. 163–88.

8 Cf. Bruno Kissoun, 'Fortifications des îles. Trois siècles d'architecture militaire en Guadeloupe: XVIIe–XIXe siècle', *Bulletin Monumental* 163:4 (2005), 343–56.

9 There are several maps of the French Antilles; the one by Jacques-Nicolas Bellin, *Carte réduite des isles de la Guadeloupe, Marie-Galante et les Saintes* (Paris: Dépôt des cartes plans et journaux de la marine, 1759), shows both fortresses near the village of Baillif and the town of Basse-Terre.

10 CAOM COL C8B 1, No. 22: Mémoire sur les fortifications de Martinique, fol. 2r: 'Ce n'est proprement, comme il dit, qu'une prison car les huit aigles qu'on a adjousté a la premiere fabrique qui n'estoit qu'une tour quarée sont si petits, si mal entendus et pour dire le vray si ridicules, qu'ils ne font paroistre ce bastiment qu'une lourde

masse de pierre entassée sans dessein n'y sans ordre, si bien qu'il ne faut faire nul cas de cet ouvrage pour la deffence de l'isle'; see also CAOM 08DFC/26 (Guadeloupe), No. 3: La Guadeloupe, tiré du Memoire sur Blondel, fol. 1r.

11 CAOM 08DFC/26, No. 2: Avis, fol. 1r.

12 P. Jean-Baptiste Du Tertre, *Histoire générale des Antilles Habitées par les François divisée en deux tomes, Et enrichie de Cartes & de Figures*, 2 vols (Paris: Thomas Iolly, 1667).

13 Ibid., vol. 1, p. 445: 'M. Hoüel ravy de cette association, appliqua tous ses soins à faire bastir une maison proche de la principale rade de la Basse-terre: il la fit d'une structure toute nouvelle à 4 faces, et à quatre étages. Dans chaque étage il y a quatre chamber de plein-pied, les murs sont de tres-belles pierres [...]. Au bas de cette maison du costé de la Mer, il y a une batterie de six pieces de canon, qui commandent la rade, et qui la peuvent deffendre de l'abord des Vaisseaux.'

14 Ibid.

15 Ibid., pp. 460–5.

16 Ibid., pp. 465–9.

17 Ibid., pp. 471–2.

18 CAOM COL C8B 1, No. 22: Mémoire sur les fortifications de Martinique, fol. 2v: 'Et au lieu du nombre des travaux qu'il y juge necessaire, [...] avec cette petite reparations qui se pourra faire pour huit mil livres on rendra cette maison qu'ils appellent de campagne une fortresse considerable et telle que quand les ennemis se seroient rendus maistres de l'isle ils n'en oseroient entre prendre le siege.'

19 Kissoun, 'Fortifications des îles', p. 345.

20 CAOM 08DFC/26, No. 5: Mémoire sur la Défense de l'Ile de la Guadeloupe et dépendantes, par Messieurs Blondel et Du Lion, 27 January 1667, fol. 4v: 'Et si le fort de la madeleine avait été trace par une personne un peu plus intelligente, et suivant les regles de la bonne fortification, il n'y aurait rien à desires.'

21 Cf. Anthony Gerbino, *François Blondel: Architecture, Erudition, and the Scientific Revolution* (London/New York: Routledge, 2010), pp. 77–84.

22 CAOM COL C8A 6, fol. 79: Charles de la Roche-Corbon de Blénac to Pierre Hinselin, 3 August 1690, fol. 79v.

23 Kissoun, 'Fortifications des îles.', pp. 343–4: 'La question qui se pose est alors de savoir si l'on doit appliquer en Guadeloupe les dispositifs que l'on établit au meme moment sur les frontières de la France, c'est-à-dire les principes du trace, bastionné, et par eextension celle du choix du fort à priviliger d'un point de vue strategique et topographique.'

24 Cf. Daniel Nordman, *Frontières de France. De l'espace au territoire, XVIe–XIXe siècle* (Paris: Gallimard, 1998); see also Philip P. Boucher, 'The "Frontier Era" of the French Caribbean, 1620s–1690s', in Christine Daniels and Michael V. Kennedy (eds), *Negotiated Empires: Centers and Peripheries in the Americas, 1500–1820* (New York/London: Routledge, 2002), pp. 207–35.

25 Cf. David Bitterling, *L'invention du pré carré. Construction de l'espace français sous l'Ancien Régime* (Paris: Albin Michel, 2009).

26 David Bitterling, 'Marschall Vauban und die absolute Raumvorstellung', in Lars Behrisch (ed.), *Vermessen, Zählen, Berechnen: Die politische Ordnung des Raums im 18. Jahrhundert* (Frankfurt am Main/New York: Campus, 2006), pp. 65–74; according to Bitterling, Vauban used the metaphor of the *pré carré*, a technical term in the art of fortification, in a national-territorial sense that prefigured the proto-physiocratic concept he developed for his economic theory in the tax levy in his *Projet d'une dixme royale* (1707). In a famous letter to the secretary of war Louvois he wrote: 'Sérieusement, Monseigneur, le Roi devrai un peu songer à faire son pré carré [...]. C'est pourquoi, soit par traité ou par une bonne guerre, si vous m'en croyez, Monseigneur, prêchez toujours a quadrature, non plus du cercle, mais du pré; c'est une belle et bonne chose que de pouvoir tenir son fait des deux mains' (Vauban to Louvois, 20 January 1673, in Albert Rochas d'Aiglun, *Vauban. Sa famille et ses écrits, ses oisivetés et sa correspondance, analyse et extraits*, 2 vols (Geneva: Slatkine, [1910] 1972), vol. 2, p. 89.

27 Alexander Exquemelin, *Histoire des avanturiers qui se sont signalez dans les Indes, contenant ce qu'ils ont fait de plus remarquable depuis vingt années* (Paris: Jacques Le Febvre, 1686; 2nd ed. Paris: Jacques Le Febvre, 1699), vol. 1, p. 134. Later on, plantations and homesteads were usually between 100 and 200 hectares; see Pritchard, *In Search of Empire*, p. 85.

28 A similar map appears in the first edition of P. Jean-Baptiste Du Tertre, *Histoire générale des Isles de Saint-Christophe, de la Guadeloupe, de la Martinique* (Paris: Jacques Langlois, 1654), p. 68.

29 Names like that of Captain Pilote, an indigenous chief, were probably given, according to Charles de Rochefort, *Histoire naturelle et morale des iles Antilles de l'Amérique. Enrichie de plusieurs belles figures des raretez qui y sont décrites. Avec un vocabulaire caraïbe* (Rotterdam: Arnould Leers, 1658), p. 15, by the French to the Carib leaders. Rochefort probably relied on information by Pacifique de Provins, *Brieve Relation du voyage des Isles de l'Amérique* (Paris: Nicolas et Jean de la Coste, 1646), p. 8: 'Le Gouverneur de cette Isle sur les François se nomme Monsieur du Parquet [...] qui gouverne son peuple avec tout plein de satisfaction de leur part; & qui par ce moyen s'est acquis tant de crédit sur l'esprit des Sauvages de son Isle, qu'il en fait ce qu'il veut; spécialement sur celuy de leur Capitaine, qui s'appelle le Capitaine Pilote: car il m'a dit avoir tout nouvellement obtenu de ce Sauvage, qu'il feroit près de son logis, une petite maison et Chapelle, pour y loger un ou deux peres Iesuites; [...] Ce Capitaine Pilote dit un jour à Monsieur le Gouverneur (ainsi qu'il me l'a raconté) que pour l'amour qu'il portoit aux François, il vouloit que tous les enfans qui naistroient désormais des Caribes ou Sauvages, portassent des noms de François.'

30 Cf. Vincent Huyghues-Belrose, 'Le nom des lieux à la Martinique: un patrimoine identitaire menacé', *Études caribéennes* 11 (2008), 2–17; Bright, '"Removed from the Face of the Island"'; and Laurence Verrand, *La vie quotidienne des Indiens caraïbes aux Petites Antilles (XVIIe siècle)* (Paris: Karthala, 2001), p. 72. For the larger context, see Philip P. Boucher, *Cannibal Encounters: Europeans and Island Caribs, 1492–1763* (Baltimore, MD/London: Johns Hopkins University Press, 1992).

31 CAOM COL C8B 1, No. 22: Mémoire sur les fortifications, fol. 1v: 'Si le dessein que le Sr. Blondel a fait esbaucher au Carenage du Cul de Saq Royal eus testé continué et mis en deffence selon son projet, il n'auroit pas seulement cousté au Roy, plus de cinquante mil escus, Mais il auroit mesme esté inutil, car le fort estant placé ~~suivant son relation~~ comme il l'eust esté ~~suivant sa relation~~ il n'eust deffendu ny le port, ny l'entrée de la hauteur, Il ne voyoit ny l'eau ny la terre ferme, et Il s'engageoit a breuser un fossé de neuf thoises de large dans le Roc le plus vif et le plus dur qui soit au monde, Ce qui auroit causé une despence infinie' (cancellations in the original).

32 While the seigniorial system was never implemented in the French Antilles, the term *seigneurie* appears in the sources indicating at least the appearance some proprietors wished to convey as feudal or noble land holders. The seigniorial system of land ownership, however, was more typical for New France or Canada; see Pritchard, *In Search of Empire*, pp. 78–88.

33 Rochefort, *Histoire naturelle et morale*, pp. 25–6: 'Monsieur le Gouverneur fait sa demeure en un Chateau, qui n'est pas fort éloigné de la Ville [de Baillif]. Il est bâti bien solidément, à quatre face. Les coins sont munis déperons, et de redoutes, de massonnerie d'une telle épaisseur, qu'elle peut soûtenir la pesanteur de plusieurs pieces de Canon de fonte verte, qui y sont posées en batterie.'

34 For Poincy's career, see Prosper Cultru, 'Colonisation d'autrefois. Le Commandeur de Poincy à Saint-Christophe', *Revue de l'histoire des colonies françaises* (1915), 289–354; see also Roulet, *La Compagnie des îles de l'Amérique*, pp. 543–58.

35 Jean-Baptiste Labat, *Nouveau Voyage aux Isles de l'Amérique*, 2 vols (Den Haag: Husson, van Duren et al., 1724), vol. 1, p. 193: 'C'est celle de feu M. le Bailli de Poincy, ci-devant Gouveneur general des Isles. On la nommoit le Château de la montagne, parce qu'elle étoit bâtie sur une montagne à une lieüe et demi du Bourg. La situation ne pouvoit être plus belle, ni la vûe plus étendue et plus diversifiée. Le Père du Tertre en a donné un dessein dans son Histoire, qui me servit à la

reconnoître, quand j'en allai voir les restes qui ne sont plus à present qu'un amas de ruïnes au milieu de plusieurs terrasses, qui marqoient la magnificence, les richesses, et le bon goût de celui qui avoit fait construire ce bel édifice. J'y trouvai encore quelques grottes assez entieres, des bassins dont on avoit enlevé le plomb, et les reservoirs des eaux d'une fontaine, dont la source est à une demie lieüe plus haut dans la montagne.'

36 Ibid.: 'J'ai dit dans plus d'un endroit, que les richesses des Habitans consistoient dans leurs Esclaves. Ce sont leurs bras, sans lequels les terres demeureroient en frîche: car il ne faut pas songer de trouver des gens de journée comme en Europe, on ne sçait ce que c'est; il faut avoir des Esclaves, ou des Engagez, si on veut faire valoir son bien. De sorte que l'Habitant qui a un plus grand nombre d'Esclaves est le plus en état de faire une fortune considérable.'

37 Rochefort, *Histoire naturelle et morale*, p. 36: 'Mais ce beau Palais presentant à l'œil une face extremément charmante, à péne la peut on jetter ailleurs. Sa figure est présque quarrée, à trois étages bien proportionés, suivant les régles d'une exquise Architecture, qui y a employé la pierre de taille, et la brique, avec une belle symmetrie. La face, qui se presente la premiere, et qui regarde l'Orient, a au devant de son entrée un large escalier, à double rang de degrez, avec un beau parapet au dessus; et celle qui a l'aspect au Couchant, est aussi embellie d'un escalier tout pareil au premier, et d'une belle et grosse source d'eau vive, qui étant receüe dans un grand bassin, est de là conduite par des canaus sou-terrains en tus les offices.'

38 Ibid.: 'Les salles et les chambres sont bien percées; les planchers sont faits à la Françoise, de bois rouge, solide, poly, de bonne odeur, et du crû de l'Ile. Le couvert, est fait en plate forme, d'où l'on a une veüe des plus belles, et plus accomplies du monde. Les fenestrages sont disposez en bel ordre: les veües de devant sétendent le longe de l'avenuë, et percent dans de beaux vallons, plantez de Cannes de Sucre, et de Gingembres. Celles du Couchant, sont terminées par la montagne, qui n'en est éloignée qu'autant que la juste proportion le requiert, pour reliever par le riche fonds qu'elle presente, la grace et les perfections de ce Palais.'

39 Ibid., p. 37: 'C'est une chose divertissante au possible, qunad aux jours de rejouissance publique, on fait en l'Ile des feus de joye, pour les nouvelles de quelque heureux succés des armes victorieuses de sa Majesté Tres-Chrestienne. Car alors les Clairons, et les Hautbois sont ouir leur son éclartant du haut de la platte-forme de ce Palais, en telle sorte que les montagnes voisines, les côtaus et les bois qui les couvrent, retentissent à ce bruit penétrant, et formen tun aimable éco qui s'entend par toute l'Ile, et bien avant en mer. Alors on voit aussi pendre du haut de la Terrasse, et des fenestres de l'etage le plus élevé, les enseignes semées de fleurs de Lis, et les drapeaus et étendars que Monsieur le General a remportez sur les ennemis.'

40 Ibid., p. 38: 'Car de ses fenestres on voit dans la bassecourt trois machines, ou moulins propres à briser les Cannes de Sucre, qui apportent à leur maistre un profit, et une revenue assuré, et qui va du pair avec celuy des plus nobles et meilleures Seigneuries de France.'

41 Du Tertre, *Histoire générale des Antilles*, vol. 2, pp. 9–10.

42 CAOM COL C8B 1, No. 6: Inventaire après décès du bailli de Poincy, 12 April 1660.

43 Ibid., fol. 3v.

44 Cf. ibid., fol. 5r-7v, for example, Georges Fournier, *Hydrographie contenant la théorie et la pratique de toutes les parties de la navigation* (Paris: Michel Soly, 1643); Nicolas Goldman, *La nouvelle fortification* (Leiden: Elzevier, 1645); and Antoine de Ville, *Les fortifications du chevalier Antoine de Ville, contenans la manière de fortifier toute sorte de places [...] avec l'ataque et les moyens de prendre les places [...] plus la défense* (Lyon: Philippe Borde, 1640).

45 Michel-Christian Camus, 'Le général de Poincy, premier capitaliste sucrier des Antilles', *Revue française d'histoire d'outre-mer* 84 (1997), 119–25.

46 For a classic comparative analysis of the plantation economy, see Eric R. Wolf and Sidney W. Mintz, 'Haciendas and Plantations in Middle America and the Antilles', *Social and Economic Studies* 6 (1957), 399–400; cf. also Robin Blackburn, *The Making of New World Slavery: From the Baroque to the Modern, 1492–1800* (New York/

London: Verso, 1998), p. 301, who sees 'a certain structural contrast between English "bourgeois" colonization and French "Absolutist" colonization' in the Caribbean plantation economy.

47 Du Tertre, *Histoire générale des Antilles*, vol. 1, pp. 169–72; Exquemelin, *Histoire des avanturiers*, vol. 1, pp. 29–31.

48 The first quote is by Du Tertre, *Histoire générale des Antilles*, vol. 1, p. 171; the second by Exquemelin, *Histoire des avanturiers*, vol. 1, pp. 33–4.

49 Du Tertre, *Histoire générale des Antilles*, vol. 2, p. 31.

50 Ibid., vol. 1, p. 172.

51 Ibid., p. 173.

52 Ibid., pp. 174–5.

53 Ibid., p. 178.

54 Ibid., pp. 180–8.

55 Ibid., pp. 101–8; quote by ibid., p. 107.

56 Poincy to Fouquet, 16 August 1639, quoted by ibid., p. 108: 'Pour les affaires de l'Isle de la Martinique elles sont en tres-bon estat, et M. du Parquet merite de grandes loüanges pour les soins et diligences qu'il y apporte, afin que tout y aille d'ordre. Il a fait faire es habitations proche du Fort Royal: d'autres à son imitation y en ont pris, de sorte qu'ils commancent fort de s'élargir. Il y a environ sept cens hommes capables de combattre, mais s'il falloit qu'ils fussent attaquez, ils n'ont pas de poudre pour tirer chacun quatre coups. Il a fait renouveller toutes les Pallisades dudit Fort Royal: tous leurs canons sont démontez, autant vaut-il, puis que les afust ne vallent rien. Il n'y a qu'un Charpentier en toute l'Isle; et entre leurs autres necessitez, celle-là n'est pas des moindres, et à laquelle specialement vous devez pourvoir et leur envoyer quelqu'un.'

57 Parquet to Fouquet, 17 August 1639, quoted by ibid., p. 110: 'Pour ce qui est de faire une ville, comme vous me mandez de commencer, il faudroit que vous m'envoyassiez quantité de massons, briquetiers, tailleurs de pierre, faiseur de chaux, Charpentiers, Menuisiers, Serruriers, Taillandiers, Cloutiers, Couvreurs, et autres ouvriers garnis de leurs outils, et autres choses necessaires pour la construction de ladite Ville, qui ne se rencontrent pas en ce pays.'

58 Ibid., p. 111.

59 Ibid., vol. 2, p. 26f.

60 Du Tertre, *Histoire générale des Antilles*, vol. 2, p. 26.

61 Rochefort, *Histoire naturelle et morale*, p. 16: 'Ce Fort commande sur toute la rade. A un jet de pierre du logement de Monsieur le Gouverneur, est la belle Maison des Jesuites, située sur le bord d'une agreable riviere, que l'on appelle pour cette raison, *la Riviere des Iesuites*. Ce rare edifice est bâty solidement de pierres de taille et de briques, d'une structure qui contente l'oeil. Les avenuës en sont fort belles; et aus environs on voit de beaus jardins, et de vergers remplis de tout ce que les Iles produises de plus delicieus, et de plusieurs plantes, herbages, fleurs et fruits qu'on y a apportez de France. Il y a même un plan de Vigne, qui porte de bons raisins, en asses grande abondance, pour en faire du vin.'

62 The habitation La Montagne was sold to the Compagnie des Indes occidentales in 1664. After financial difficulties it was sold again two years later to the Seigneur de Clodoré, royal governor of Martinique, Betrand d'Ogeron, governor of Tortuga, Jacques Giraud, and the Sieur de Chambré. The latter part was bought by the Jesuits. In 1679, the Marquis de Maintenon, Charles François d'Angennes, was the sole proprietor of the habitation and went to his widow and her brother Giraud de Crésol. It remained in the hands of the family Crésol until 1771 and was bought eventually by the Pécoul in 1809, who remained the proprietors until 1917. Since 1918 the estate is run by the family of Depaz, who built a new mansion in 1922 after the image of the old Jesuit monastery that got destroyed together with the other original buildings of the habitation in the eruption of Mont Pelée in 1902.

63 Blondel, L'isle de la Martinique, 1670 (see Figure 1).

64 Du Tertre, *Histoire générale des Antilles*, vol. 2, p. 27.

65 Rochefort, *Histoire naturelle et morale*, p. 15: 'Monsieur le gouverneur a honoré un fort long tems cet agreable quartier de sa demeure, laquelle il faisoit en une maison qui est bâtie de briques, guéres loin de la rade, pres de la place d'armes, en un beau vallon, qui est arrosé d'une asses grosse rivière, qui tombe des montagnes. [...] Cette maison, est entourré de plusieurs beaux jardins, qui sont bordez d'arbres fruitiers, et embellys de toutes less rarétez, et curiositez du païs. Monsieur le Gouverneur a quitté cette demeure depuis environ deus ans, à cause qu'il ne se portoit pas bien en ce quartier où elle est située, et en a fait present aus Jesuites, comme aussi de plusieurs belles habitations qui en dépendant, et d'un grand nombre d'Esclaves négres qui les cultivent.'

66 Du Tertre, *Histoire générale des Antilles*, vol. 2, pp. 27–8: 'Ce grand nombre d'Esclaves noirs dont il parle, se reduit à huit ou neuf: ces beaux 'jardin bordez d'arbres fruictiers, et embellis de toutes les raretez et curiositez du pays, sont des chimeres; il n'y a autre chose que du petun et de manyocs.'

67 Rochefort, *Histoire naturelle et morale*, p. 30.

68 Père Jean-Baptiste Labat, *Nouveau Voyage aux Isles de l'Amérique*, 6 vols (Paris: Guillaume Cavelier, 1722), vol. 1, p. 74.

69 CAOM COL C8B 1, No. 22: Mémoire sur les fortifications, 1669, fol. 1v d.: 'Le fort qui a esté fait au bourg St. Pierre a esté tres mal placé, et ce luy qui l'a fait bastir ne connoisait pas les scituation qui donnent ou qui ostent l'avantage, Il es sur le bord de la mer, fort esloginé du Mouillage et commandé d'une hauteur voisine qui voir jusqu'au talous les hommes qui sont dans ses murailles, Ainsy on peut dire qu s'il ne servoit de prison pour mettre les Criminels, c'est ouvrage seroit inutil.'

70 Ibid., fol. 2r: 'Il est necessaire que le cul de sac Royal soit fortifiée et que la redoutte proposée soit faitte, et jusqu'à ce que tous ces ouvrages soient achevés nul ne se peut dire Maistre de la Martinique, car les habitans ne seront jamais ny obeissans ny disciplinable, s'ils n'ont un frain qui les fasse craindre.'

71 Labat, *Nouveau Voyage*, vol. 1, pp. 196–9.

2

Ambitions to empire in India: Pondichéry as an imperial city in the Mughal state system

The materiality of building projects reveals more about the different allegiances of a colony than can be seen just from looking at the expressed ideas of the colonizers. In the following case local material culture was present in all the projects the French undertook. The architecture of Pondichéry at the Indian Coromandel Coast was supposed not only to represent the French empire, but also to establish the settlement as an imperial city in the Indian state system. Buildings represented the material cultures of France, India, and Tamil people not only in the use of local and colonial actors, indigenous building material, and hybrid elements of style, but also in the way the people reacted emotionally to the new constructions. They functioned as affective buildings that served the establishment of a hybrid emotional community in Pondichéry.

The French commitment to empire is perhaps best represented by the monumental design of the Porte Royale of Fort Louis in Pondichéry.[1] Compared to the structures in the Antilles the entrance gate to the new fortress, planned by the royal engineer de Nyon in 1705 or 1709, was unsurpassed (see Figure 12). Its tripartite façade, with two triangular gables and a central curved pediment, was decorated with emblems similar to the more representative gates in America. Since the Compagnie des Indes orientales that Colbert founded in 1664 was still proprietor of Pondichéry at this time, its coat of arms, a single Fleur-de-Lys with the motto *Florebo quocumque ferar* ('I flourish wherever I am carried'), appeared above the entrance to the left. The gate to the right showed the crest of the current governor, at that time Guillaume-André de Hébert. The middle segment displayed the royal coat of arms below a large clock and in the gable a sun with the personal device of Louis XIV 'Nec pluribus impar'. A cupola resting on an octagonal turret crowned this stately gate.

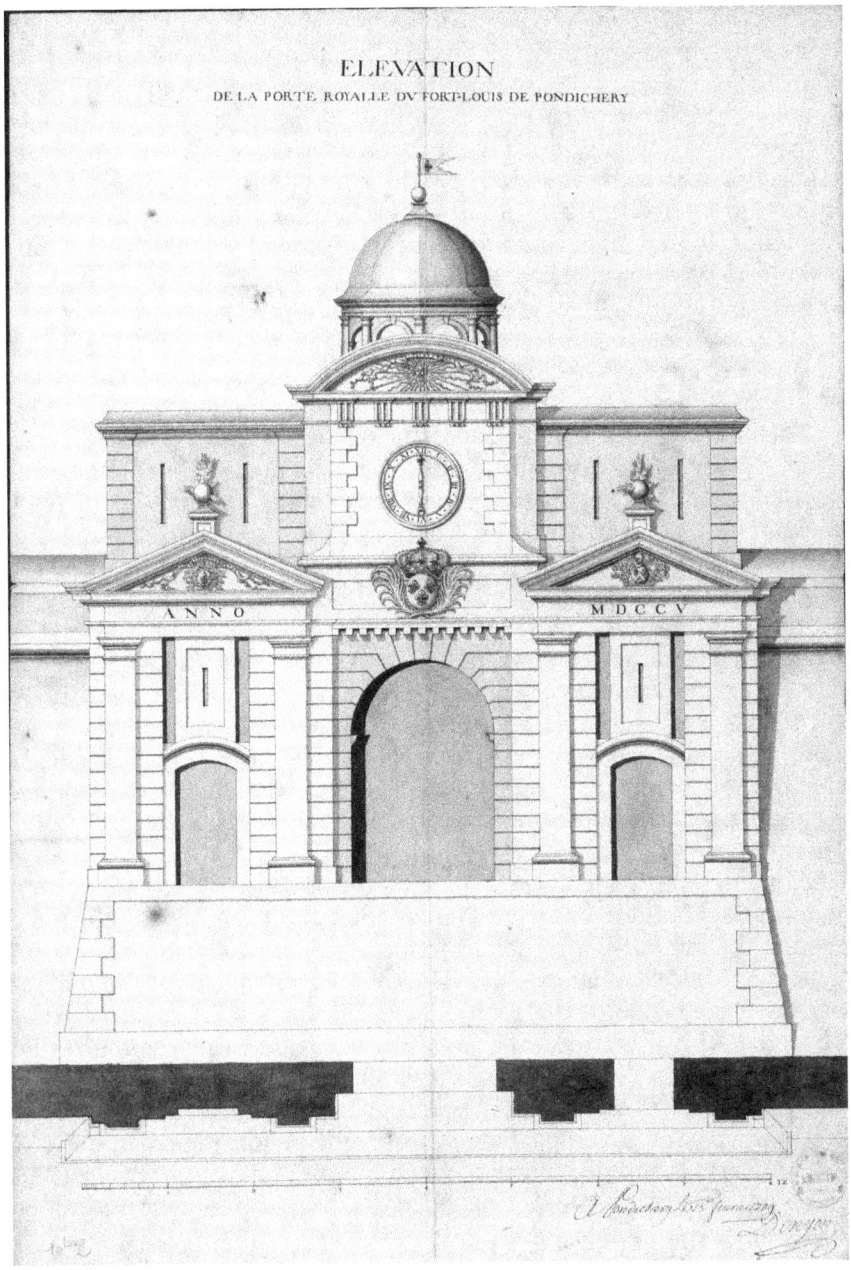

12 Nyon: Elevation de la Porte Royalle du Fort Louis de Pondichery, dated 15 February 1709; the date on the façade, however, reads 1705 when the governor François Martin was still incumbent (CAOM 26DFC/10terC).

Together with the slightly less monumental Porte Dauphine the Porte Royale was part of a pentagonal-shaped fortress in the style of Vauban's citadels in Tournai, Lille, and Douai (see Figure 13). As in Flanders, the application of the ideals of modern fortification may be seen in Pondichéry as portraying the formation of a French territory on occupied foreign ground. The old fortress, built according to plans drawn up by a Capuchin monk called Père Louis during the tenure of French India's first governor François Martin, was slighted, while the old town remained more or less intact. A plan dating from 1704 shows the new fortress overlapping the old one, surrounded by the domain granted to the East India Company in 1675 by the Sultan of Bijápúr, later on by Sher Khan Lodi and Rám Rájá, ruler of the Maráthás.[2]

The new city was planned according to a geometrical grid structure.[3] This, however, was not originally a French plan. The scheme was devised during the Dutch occupation of Pondichéry from 1693 to 1699. Rectangular blocks served as space for residential and commercial buildings, bringing order to the unstructured agglomeration of the original town.[4] Later on, after the Dutch left the city in 1699 having received 16,000 pagodas as recompense for the fortification works they had carried out, the reinstated governor Martin continued working on the city improvements. The workforce available to Martin in these precarious years of French presence in India was extremely small. Of the sixty soldier-workmen available before the Dutch occupation, only thirty-four now remained and probably had help from indigenous workers to build the fortress, houses, magazines, and stores according to the Dutch plan.[5]

Around 1700, Pondichéry remained predominantly an Indian city. A few brick buildings for the French company agents were clustered around the old fortress in no apparent order. Martin occupied perhaps a 'fine', but also a small palace.[6] The thatch-roof houses of the Indians occupied the space to the west. These buildings were made of wood plastered with a white chalk-based composition called *chunam* that was made of seashells. Indian and French houses may have looked, therefore, not too dissimilar, at least concerning their materials. Later on, however, the governor Pierre-Christophe Lenoir issued an order to use only bricks and tiles for the French buildings. But this was in 1730, implying that such an order was necessary to impose a more French style on the city's appearance.[7]

In fact, Fort Louis, the enclosure of the whole city, and the monumental projects devised by de Nyon in 1704 awaited their completion for many years (see Figure 14). We should not be misled by the apparent growth of buildings in the city insinuated by the series of charts and plans.

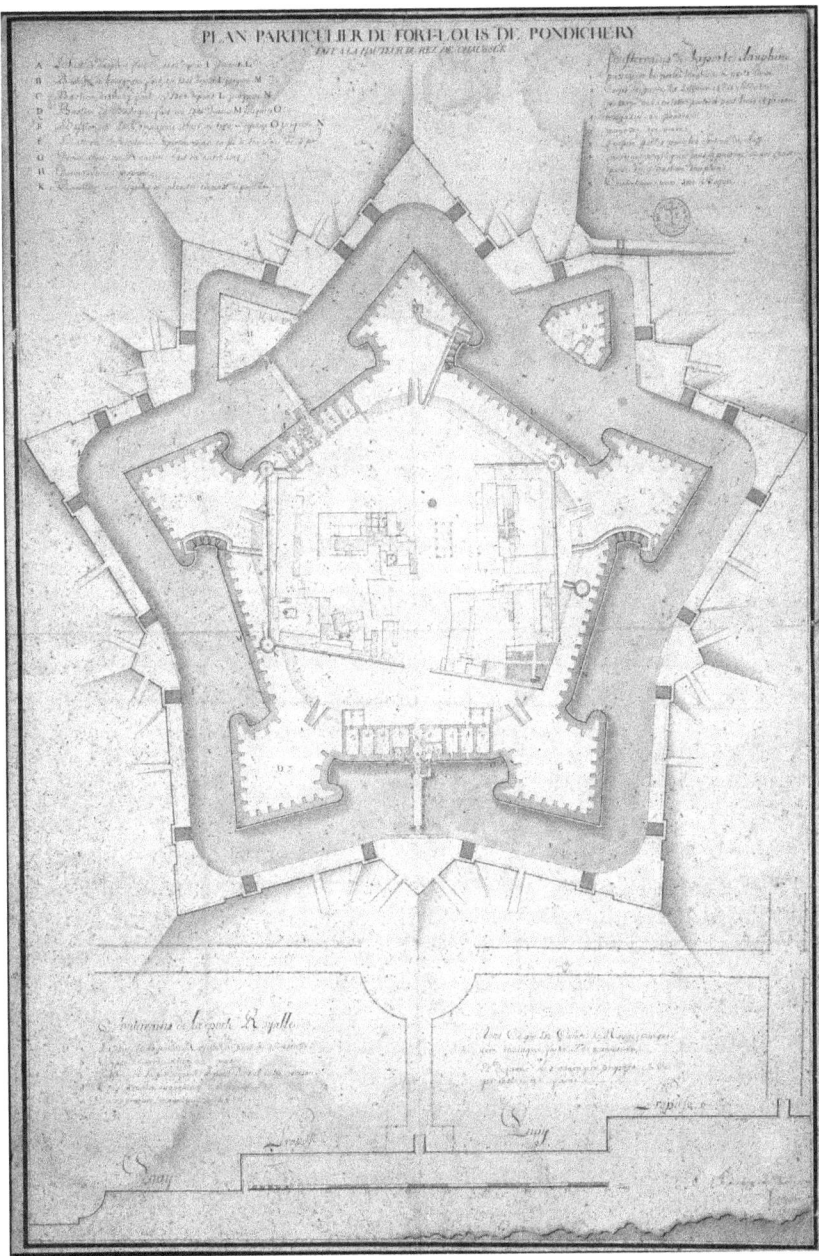

13 Nyon: Plan particulier du fort Louis de Pondichéry, fait à la hauteur du rez-de-chaussée, 15 February 1709 (CAOM 26DFC/10A). The new fortress with the Vauban-style layout can be discerned from the outline of the older castle.

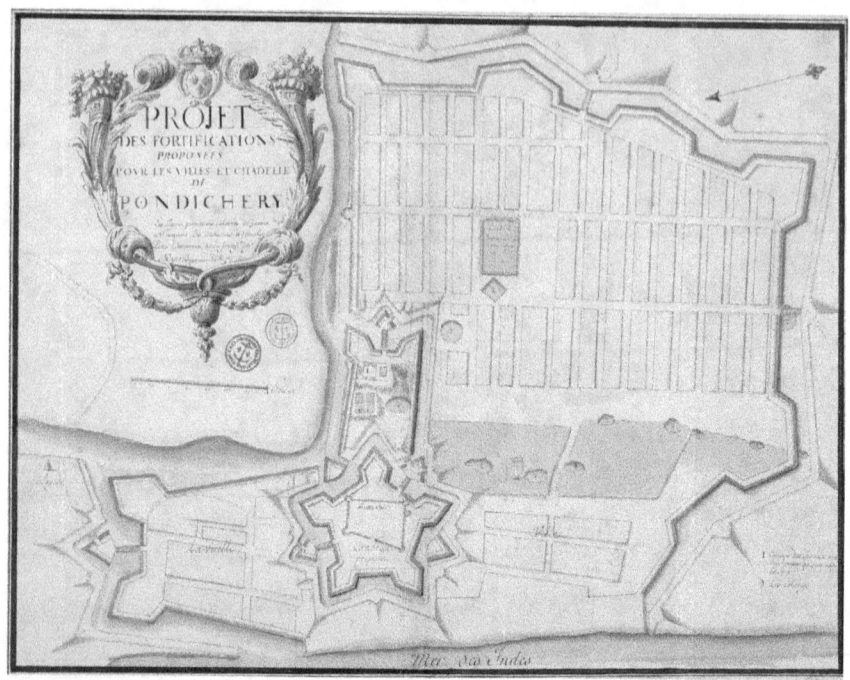

14 Nyon: Projet des fortifications proposées pour les ville et citadelle de
Pondichéry, ca. 1700 (CAOM 26DFC/4B).

French maps from 1700, 1704, 1714, 1721, and 1733 give the impression
of a continuous expansion, but hide the fact that the development of
the city progressed only slowly.[8] It was only through the increasing
stability of the India Company in the decades following the financial
collapse in France in 1720 that ships carrying material, merchandise,
and funds arrived with almost regular frequency in Pondichéry.[9]

It was the financial funds that were of great importance for the
French, however, in order to gain credibility with the much wealthier
Indian traders. Due to the disaster caused by the failure of the speculative
scheme of John Law, the India Company was highly indebted, not only
in Europe but in India too. In 1722, when for the second year no French
fleets had arrived, the governor Lenoir had to depend on the company's
credit with the rich natives. 'It was by their [the rich natives'] aid and
forbearance alone', Malleson writes, 'that Lenoir was able to save the
credit of the colony in this period of dire necessity.'[10]

It seemed paramount, therefore, to reassure the trust of the local
commercial elite in French future solvency. Heavy investments were

thus put into the building effort in Pondichéry. The construction of Fort Louis, however, which appeared in its ideal form on paper in a sketch by de Nyon as early as 1702, took some time. The pentagon shape seems to have been completed between 1704 and 1714 with the walls of the old fortress still remaining within the interior court of the new fortress. We can assume that the Porte Royale, facing east to the sea, was finished then, too. The entire sea front, however, was still completely open. There was neither a paved quay wall nor a proper defence system. A small bonnet in front of the citadel seemed insufficient, but was not razed and replaced before 1743.[11]

Despite the poor reputation of the two governors d'Hébert and Dulivier, detailed plans of the city from 1707 to 1718 show some improvements. A large geometrical garden was built for the Jesuits next to the houses of the indigenous Brahman elite of the city (see Figure 15). The garden had several ponds for purification rites of the Hindu population, and adjacent to the Capuchin convent stood a church next to a pagoda. In 1722, under the acting governor Lenoir, a more elaborate church was built inside Fort Louis (see Figure 16).[12] Its classical façade had similar ornamental features and was placed on the same central axis as the Porte Royale.[13] The architecture thus helped to achieve the dramatic revelation of the most elaborate French buildings upon entering the fortress.

When a delegation of the Indian nobility, among them the Nawab Asseraly-Khân, his brother-in-law Chander-Saeb, the Grand Diwan of Arcat, and other dignitaries, visited Pondichéry on 31 August 1740 they were 'ecstatic', the author of a memorandum wrote, to see the order and state of the place – a city with an air very unlike that of others. They were happy to have chosen Fort Louis for asylum and not less astonished to see the form of the citadel. 'It is true', the memorandum concludes, 'that the eyes of the Moors are not yet adapted to the European fortification.'[14]

Three main city gates, one to the north, the Porte Madras, and two to the west, the Porte Villenour and the Porte Vadaour, were designed around 1733 as monumental and representative landmarks. The Porte Madras looked oversized in proportion showing only a small gate flanked by a massive blind wall, divided by four columns and a frieze ornamented with helmets and shields (see Figure 17). The exterior of the Porte Villenour, on the other hand, was less decorated but as solid as the Madras gate. The gates were all part of the city walls comprising the old and new town as well as the southern extension. Since the commencement of the government of Joseph-François Dupleix in 1742, over sixteen bastions had been added including the seaward defences.[15]

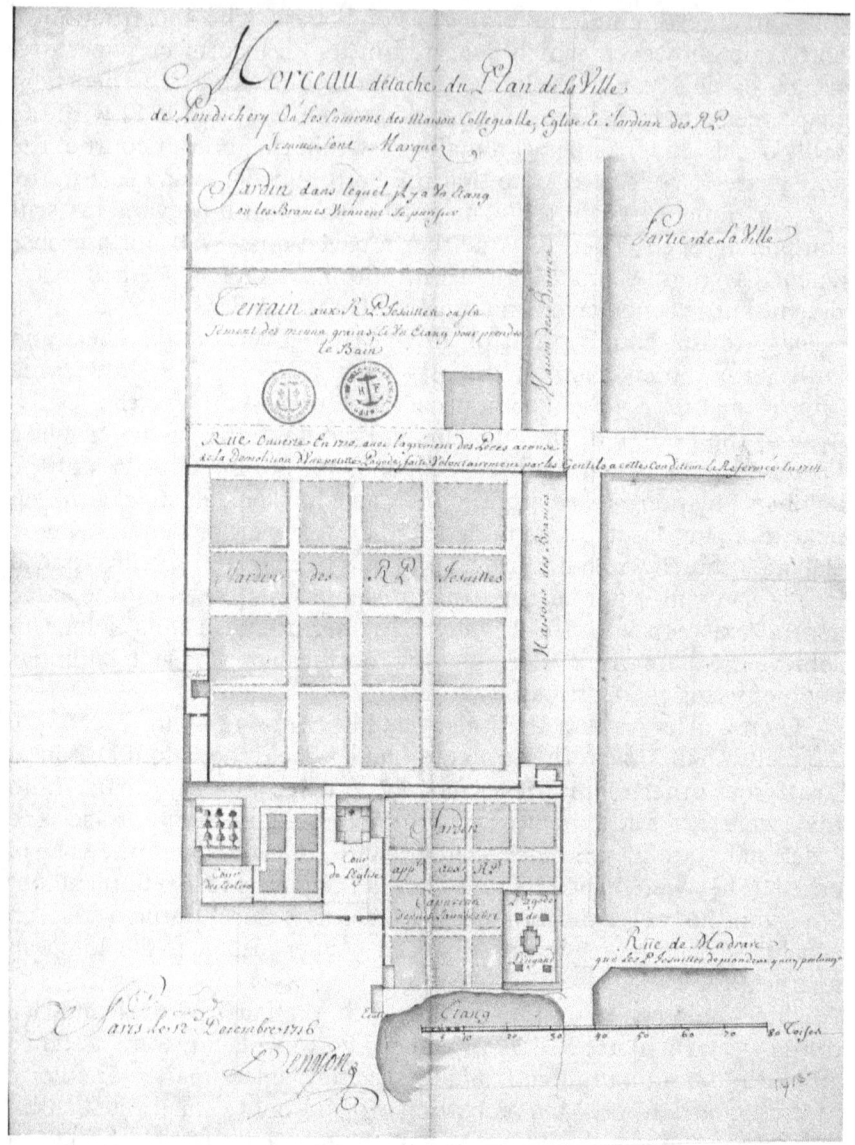

15 Nyon: Morceau détaché du plan de la ville de Pondichéry, ou les environs des maison collégiale, église et jardins des R. P. Jésuites sont marquez, 12 December 1716 (CAOM 26DFC/14B).

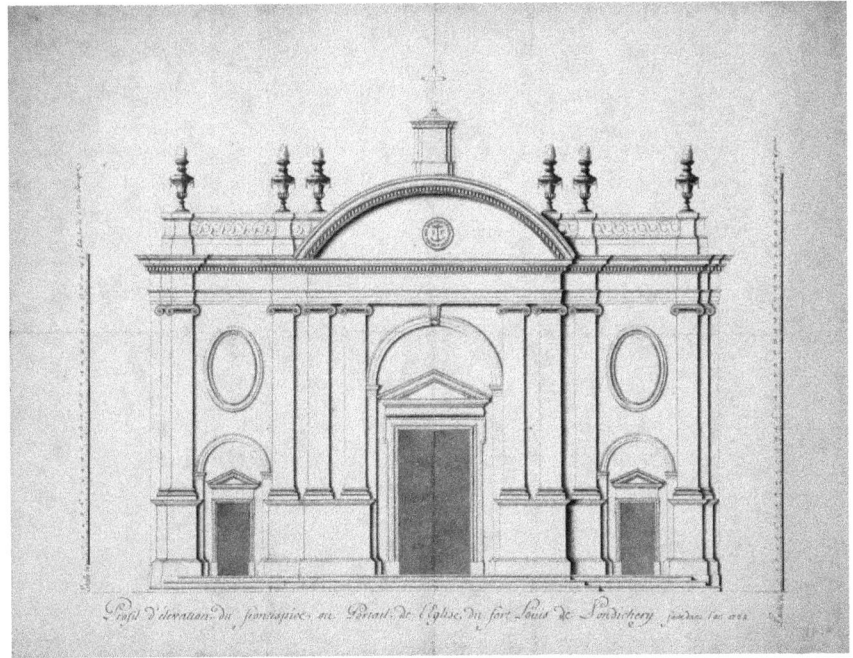

16 Profil d'elevation du frontispice ou Portail de l'eglise du fort Louis de
Pondichéry, 1722 (CAOM 26DFC/18B).

The governor Dupleix is known as the most illustrious administrator
in the history of French India.[16] Together with Bertrand François Mahé
de La Bourdonnais and Charles Joseph Patissier de Bussy-Castelneau
he is celebrated in colonial hagiography as having defended and con-
solidated the colonial empire in the East throughout the two decades
preceding the Seven Years' War. In 1748, Dupleix successfully held
Pondichéry against a British siege. During his tenure the city flourished
and relations with the neighbouring Indian principalities were at their
height. The governor also held the title of an Indian Nawab, which
meant that his authority over Pondichéry and its territory was recognized
by the Mughal emperor himself.[17]

Underlining this personal ambition to rulership, but also to French
imperial rule in India, several building projects were pursued under
Dupleix. Inside Fort Louis the construction of a new government palace
started in 1738.[18] When it was completed in 1752, it was perhaps the
most impressive edifice in the French empire (see Figure 18). Its façade
was 80 metres long, adorned with twenty-four columns, crowned with
three pediments, and topped by a balustrade. The marks of authority

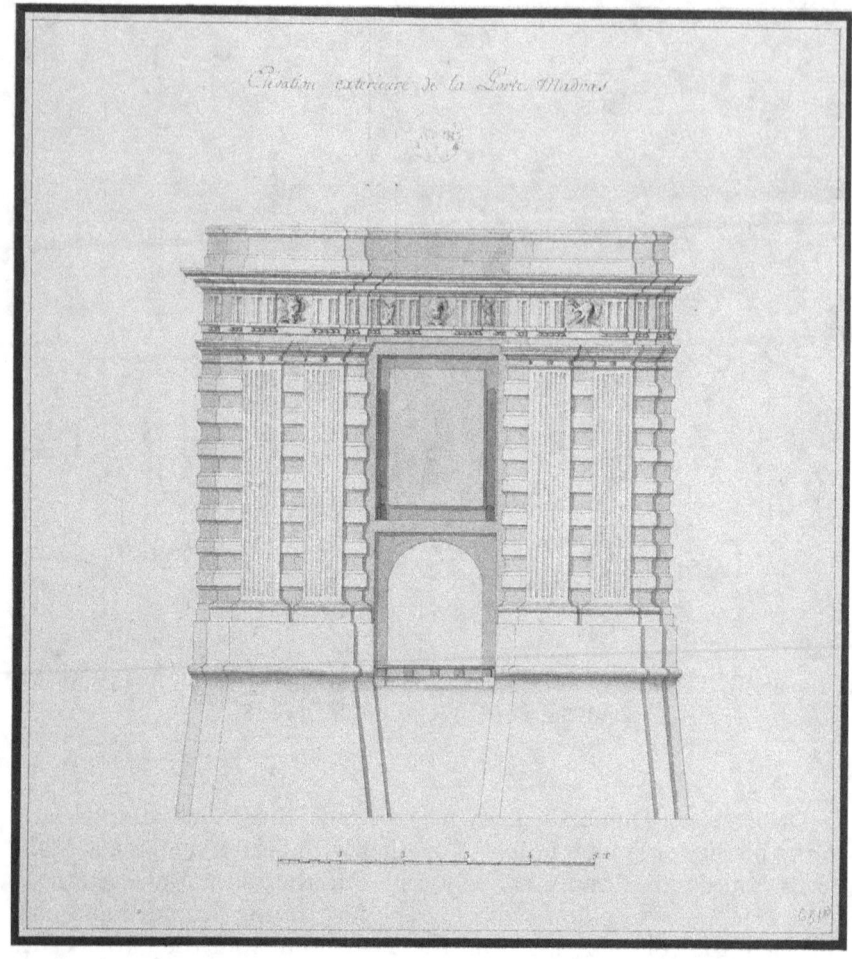

17 Elévation extérieure de la porte Madras, 1788 (CAOM 26DFC631B).

were of the same order as on the Porte Royale: in the left gable there was the emblem of the India Company, in the centre the royal coat of arms with the Fleurs-de-Lys, crown, spoils of war, etc., and within the right pediment the crest of Dupleix (two fishes above a pyramid and a star). The layouts of the palace by Champia de Fontbrun, an infantry officer, reveal an impressive amount of material and architectural splendour. Inside the palace a large entrance hall led to a staircase to the upper floor where the wall decoration and the stuccoed ceiling could easily compare with Parisian standards.

[56]

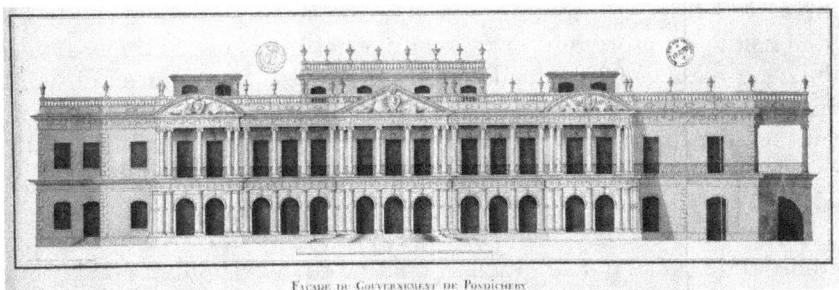

FACADE DU GOUVERNEMENT DE PONDICHERY

18   Champia de Fontbrun: Façades du gouvernement de Pondichery du côte de l'entrée, 1755 (CAOM 26DFC/78C).

Dubash Ananda Ranga Pillai, the Tamil chief translator of Dupleix, who was himself a powerful local dignitary of Pondichéry, described the new building in his famous diary.[19] He embedded the description in a meeting that took place between him and the governor, underscoring the importance of representing French power through style towards the indigenous princes:

Later the Governor sent for me, and related what Murtazâ 'Ali Khân [a Nawab from Vellore] had written about the peace between Salabat Jang and the Nânâ, but complained that nothing was said about Salabat Jang's being accompanied by 25,000 Maratha horse under Mulhari Râo or his sending for Saiyid Lashkar Khân; either it had been forgotten or the interpreters had left it out. I thus paid him my compliments, saying, 'You enjoy the good fortune of kings. As you sent an army to help Salabat Jang, what wonder that they overthrew the Marathas? You will assuredly conquer the Pâdshâh himself and sit upon his throne, and at the sound of your name, Pâdshâh, viziers, and nobles, Mussalman and Hindu alike, all tremble. Your glory dazzles like a million suns.' He was delighted with my words. Afterwards taking me up to the first story of the *Gouvernement*, he showed the hall plated with silver, containing the great mirror that has come from Europe and with windows hung with green velvet curtains fringed with lace. 'Is not this fine?' he exclaimed. I replied, 'Sir, the longer I behold the *Gouvernment* the greater is my wonder. Allbeit such a palace is but worthy of you.' Hearing my words with great joy, he continued to speak about it. We then talked of other matters. At last, I took leave and came away.[20]

The scene immediately conveys a sense of the importance of outward appearance, how Ananda Ranga Pillai grasped Dupleix's ambition to assume a powerful position among Indian nobles, and what words of praise he should choose in order to please this ambition. The hall in the bel étage of the palace is a fitting setting to back Dupleix's pretensions of glory and even an imperial mandate. Perhaps Ranga Pillai exaggerates his accolade intentionally to please a rather difficult character to deal with. The tone of the dubash's language is deferent, but the distance of the author's account allows an assumuption that he well understood how to handle and influence Dupleix to his advantage.

The *Gouvernement* was a fitting expression of the French imperial ambition in India, not so much in relation to metropolitan France, but rather in the direction of the Mughal emperor in Delhi.[21] Dupleix and his predecessors knew about the importance of this authority to the rulers in the Deccan and this part of South India. The architecture was intended to awe the visitor and inform him of the titular superiority of the French governor. Thus Dumont, in his sketch of the cross section showing parts of the interior of the palace, places a herald and a noble dignitary in a palanquin carried by four servants in front of the entrance guarded by a French soldier (see Figure 19).

The style, however, Dupleix adopted for his display of power was not, as the designs of the buildings imply, entirely colonial in the sense that it stood in opposition to native variations. It is well known that the governor dressed himself in oriental fashion during diplomatic receptions held in splendidly adorned tents.[22] Indian dignitaries in Pondichéry merged European features with the local

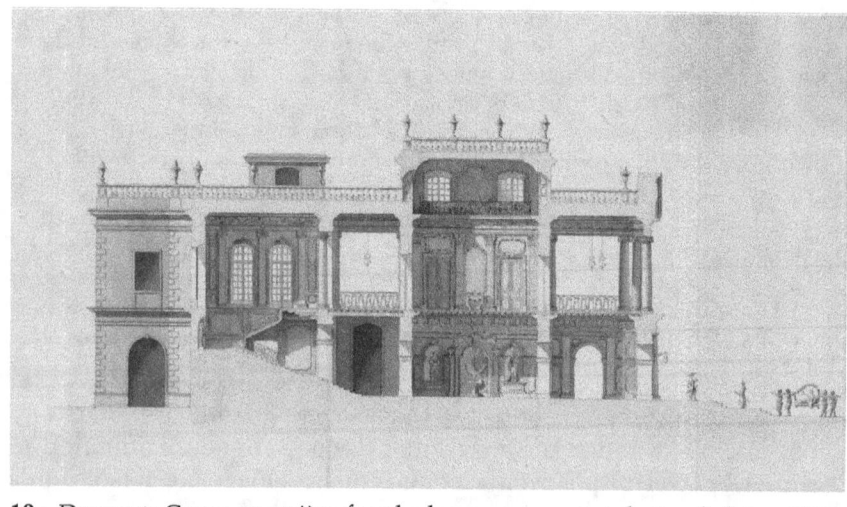

19 Dumont: Coupe et arrière-façade du gouvernement de Pondichéry, 1755, Detail (CAOM 26DFC/85A).

Tamil building tradition. Ranga Pillai, for example, adapted French architectural style for his town house (built 1735). His mansion had a patio with a gallery on the first floor that was adorned with Doric columns and a wrought-iron balustrade, and Tamil-style columns in the adjacent room on the ground floor. Like the dubash, his house represented the aesthetic features of two elite cultures and thus demonstrates the close relationship between colonial ambition and Creole appropriation.

The building practices for large projects in Pondichéry such as these were also based on this common ground between the Tamil and French societies. Concerning the material, most of the buildings were built of bricks; only a few were made of dressed stone (*pierre de taille*). As was the case in the Antilles, the defence works at Pondichéry were a permanent construction site. In 1740, for example, repairs had to be conducted at the Porte Dauphine. Since the humidity had damaged the walls, they had to be rebuilt using a dry masonry technique, which required them to be dried in the sun for a period of three months. The wooden beams and planks of the Porte Royale had to be replaced, too, and a new brick staircase was built to replace its old ladder, used to reach the cupola containing the bell and clock. The brick walls of both gates were painted ochre-red to simulate a massive stone façade. Similar repairs and alterations were made to the city defences. The Porte Villenour, for example, was decorated with a new frontispiece bearing large numerals for the construction year as well as the coats of arms of the controller-general of finances and the governor-general of Pondichéry. Several thousand trees were planted along the streets.[23]

Bricks and lime for mortar were produced on site in large brick- and limekilns situated to the south-west outside the city walls.[24] Iron, lead, and other materials such as palms, straw, baskets, bamboo, etc. were either produced near Pondichéry or imported from places such as Ceylon. Only commodities like furniture and mirrors were transported from Europe.

The destruction of Pondichéry in 1761 by British forces ended the golden age of French India and nearly all of France's imperial dreams. Dupleix, who had fallen into disgrace several years before, and had long fought for his juridical vindication, witnessed helplessly from far away in Paris the end of his project. New plans to rebuild the city emerged, however, shortly after Pondichéry returned into French possession as part of the 1763 peace treaty with Britain. And the planning produced several documents that give a better insight into the agency involved with the construction. The new chief engineer of Pondichéry, Jean Bourcet, began writing memos on the subject of reconstruction in 1765. Most important for Bourcet and for the India Company was, of course, the question of costs. Estimates for the total cost of rebuilding the city walls (Fort Louis was not going to be re-established) ranged up to one and a half million livres tournois.[25]

But next to costs, Bourcet emphasized the importance of continuing the accustomed practices in the royal works, which alone are good and free of 'shenanigans', that is, corruption.[26] He intended to build the new fortifications in four 'campaigns' that would cost much less than their equivalents in Europe, but came at huge expense with no increase in quality.[27] Bourcet's memorandum of 1765 reveals a particular sense for precise planning. The catastrophe of the city's destruction offered an opportunity that all planners in the modern age seemed to cherish: a tabula rasa that allowed a fundamentally new general plan for the urban space to be developed. Capable people, provided with proper funds, were to be employed and efficiently controlled by a team of administrators consisting of the governor, the two first engineers, and the contractor, who were themselves checked scrupulously by the Company's experts.[28]

Concerning the work itself, however, Bourcet relied on indigenous expertise. He consulted with several Indian master masons in order to get information about the price for bricks, the costs of transport, the ideal measure for the chalk, and how much the daily costs would be for an Indian mason accompanied by his sub-workers.[29] The Indian experts – masons, carpenters, and blacksmiths – were divided into three classes, who were paid according to the regulations of Karaikal, a municipality to the south of Pondichéry. Additional payment was made for indentured labourers (coulis), women, and girls (biches), but at such low rates that continual complaints led Bourcet to raise their salaries by a moderate amount.[30]

For bricks, tiles, chalk, and wood for construction, Bourcet contracted an Indian supplier called Ramalingant. But the chief engineer complained about the bad quality of the hollow tiles, which a second controller, the engineer Abeille, attributed to the gravelly earth used for its production.[31] Ramalingant, therefore, received detailed instructions on how to produce the material needed for the construction site. Bricks were to be formed according to a model provided by the chief engineer, only with the best earth and well fired, and transported to the city at the cost of the contractor. The French ordered material from Ramalingant for a total sum of 235 pagodas that was equal to 1,880 livres tournois.[32] In 1768, Bourcet proudly presented the first results in the form of a plan and report to the Company. He emphasized the low cost of the works, which he had accomplished despite the 'inexperience of the Negroes' (*l'inexpérience des noirs*) and the small number of Europeans.[33]

The racism expressed in the language of Bourcet's report was also reflected in the ordering of the new city. His colleague Abeille mentions the plan for a canal that was necessary to improve hygiene and the climate, but would also separate the town into a 'white' and a 'black' quarter.[34] But the hope of introducing more order to the city by this means was to be disappointed. Later plans show that the canal was only partially completed.[35] The appearance of individual buildings, the private residences of Indian and European citizens, might have been more homogeneous than the separation of a black and white town implies. A list from 1765 mentions the individual private residential and commercial buildings, distinguishing European and Indian proprietors, large and small sizes, and the material used for the rooftops. The majority of buildings then were in local hands, and of 1,959 'Malabar' houses, 122 were built of stone, twice the number of the 65 European stone houses.[36] Post-war Pondichéry thus must have looked more like a Creole city than the planner's ideal of a regular colonial town.

All of these accomplishments of engineering, logistics, and budgeting came at the cost of the exploitation of a low-paid indentured Indian workforce and could not hide the fact that the imperial ambition missed its moment. The new government building designed by Bourcet was not only smaller than the old one, but also less representative. It stood not in the centre of the town at the location of the old fortress, but was inconspicuously situated in a side street. It is perhaps characteristic that the Company's storehouses and offices occupied the centre in his master plan (see Figure 20). The space of the former imperial centre of French India was vacated. Instead, as Abeille described it to the Company directors in a memorandum and its accompanying plan of the new centre, only the buildings necessary for commerce were to be constructed: the archives, treasury, property, and offices. A large Hotel de la Compagnie dominated the new ensemble, housing secretaries, clerks, and cashiers.[37]

In French India the rule of the Company began when the imperial and military ambitions of individuals like Dupleix or Bussy ultimately failed. Fort Louis with its great gates and the government palace, the colonial symbols of global empire, were gone, but the Creole city with its 'white' and 'black' quarters remained.[38] A new form of empire, one of racial segregation and a capitalist regime, showed its face in Pondichéry, and would continue to dominate the picture until the French departed from India in 1954.[39]

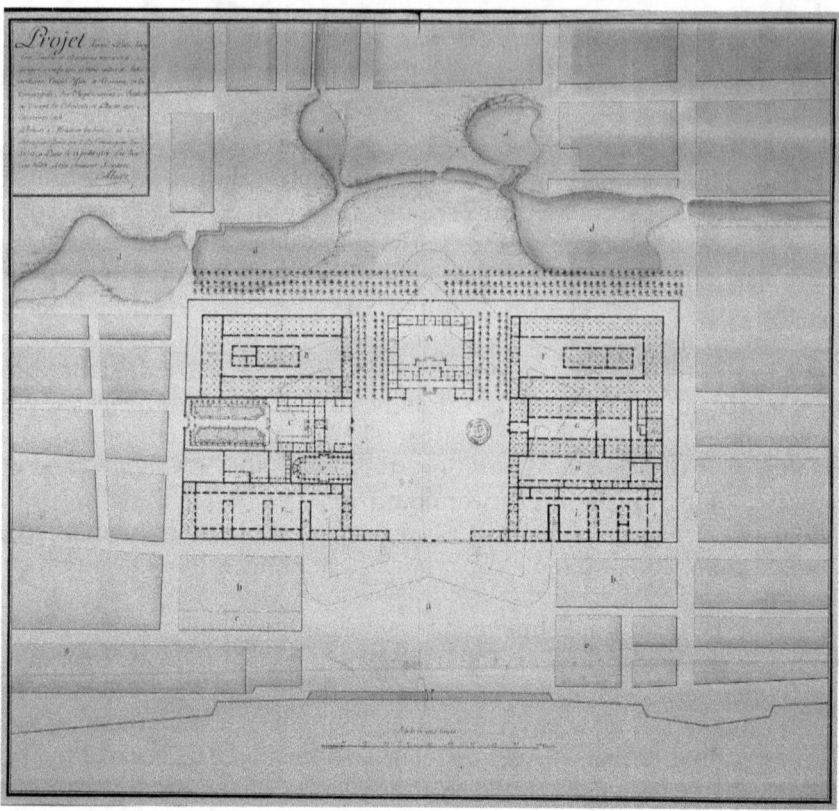

20 Abeille: Projet de bâtiments propres à renfermer les archives, trésors, effets et travaux de la Compagnie, 28 July 1768 (CAOM 26DFC/224A).

Notes

1 For the early modern history of French India, see George Bruce Malleson, *History of the French in India: From the Founding of Pondichery in 1674 to the Capture of That Place in 1761*, 2nd ed. (Edinburgh: John Grant, 1909); Jacques Weber, *Compagnies et Comptoirs. L'Inde des Français, XVII–XXème siècles* (Paris: Société française d'Outre-Mer, 1991); Jacques Weber, *Les relations entre l'Inde et la France, de 1673 à nos jours* (Paris: Les Indes savantes, 2002); Philippe Haudrère, *Les Français dans l'océan Indien (XVIIe–XIXe siècle)* (Rennes: Presses universitaire de Rennes, 2014); for the French East India Company, see Philippe Haudrère and Gérard Le Bouëdec, *Les Compagnies des Indes. XVIIe-XVIIIe siècles* (Rennes: Presses universitaires de Rennes, 2011); for the early period, see Glenn J. Ames, *Colbert, Mercantilism, and the French Quest for Asian Trade* (DeKalb, IL: Northern Illinois University Press, 1996); Paul Kaeppelin, *La compagnie des Indes Orientales et François Martin. Etude sur l'histoire du commerce et des etablissements français dans l'Inde sous Louis XIV (1664–1719)* (Paris: A. Challamel, 1908; repr. New York, 1967).

2 Malleson, *History of the French in India*, p. 27.

3 For an account of Pondichéry's architectural urban history in view of city plans, see Jean Deloche, *Le vieux Pondichéry (1673–1824) revisité d'après les plans anciens* (Pondichéry: Institut français de Pondichéry, 2005); see also Jean Deloche, *Pondicherry Past and Present: Pondichéry hier et aujourd'hui*, 2nd ed. (Pondichéry: Institut français de Pondichéry, 2019).

4 AN Den Haag, VEL 1098: Plan van fodres en stad Pudichery, 1694.

5 Malleson, *History of the French in India*, p. 27.

6 Ibid., p. 37.

7 See Haudrère and Le Bouëdec, *Les Compagnies des Indes*, p. 47.

8 See the collection of city maps in CAOM 26DFC/4B, Projet des fortifications proposées pour les ville et citadelle de Pondichéry, [1700]; 26DFC/7bisA, Plan général des dépendances de Pondichéry, aux Indes orientales, sur la côte de Coromandel, avec les ouvrages proposez et faits en 1702 et 1703, 9 February 1704; 26DFC/11A, Carte général des villes, forts et despendances de Pondichéry, sur la coste de Coromandel, avec les nouvelles acquisitions faites depuis l'année 1707, 1714; 26DFC/15B, Plan du projet des fortifications proposées des villes hautes et basses de Pontichéry, 1721; 26DFC/20bisB, Plan des ville et citadelle de Pondichéry, 1733.

9 Malleson, *History of the French in India*, p. 58.

10 Ibid.

11 Ibid., p. 98.

12 The bell tower was rather narrow and only slightly higher than the top of the façade (see 26DFC/17B).

13 CAOM 26DFC/19B, Plan du fort de Pondichéry au rez-de-chaussée, ca. 1723.

14 CAOM 26DFC/98, Cossigny to Saint-Martin, commander at the Île de France, Pondichéry, 1 October 1740, fol. 4v f.: 'En sorte que ce supplement donne aujourd'huy a nôtre ville un air bien different de celui quelle avoit cy-devant. Les Maures qui se trouvoient icy lors de ce travail, le Nabab, Asseraly-Kan, son beau frère Chander-Saeb, le Grand Divan d'Arcat, Cambadour designé Nabab, et quantité de seigneurs de leur suite, qui firent leur entrée le 31 d'aoust furent extasié de voir l'ordre et l'etat de cette place, se felicitant fort de l'avoir choisis pour leur azile ou leurs femmes son encore, ils ne furent pas moins étonné de voir la forme de nôtre Citadelle, il est vray que les yeux des Maures ne sont point trops faits aux fortification européenne.'

15 Malleson, *History of the French in India*, p. 216.

16 See Marc Vigié, *Dupleix* (Paris: Fayard, 1993); Alfred Martineau, *Dupleix et l'Inde Française*, 4 vols (Paris: Champion, 1920–8). For a critical account of colonial historiography, see Massimiliano Vaghi, 'Alfred Martineau et la « genèse » du protectorat. Le cas indien (1745–1761)', *French Colonial History* 14 (2013), 71–88. On Mahé de La Bourdonnais, see Philippe Haudrère, *La Bourdonnais: Marin et aventurier* (Paris: Desjonquères Editions, 2013); Pierre Crépin, *Mahé de La Bourdonnais, gouverneur général des Îles de France et de Bourbon, 1699–1753* (Paris: Leroux, 1922).

17 For the Mughal state system in India, see Muzaffar Alam and Sanjay Subrahmanyam (eds), *The Mughal State* (Oxford: Oxford University Press, 1997).

18 CAOM 26DFC/22B, Gerbaud: Plan du premier, ou bel étage, du gouvernement projeté, 15 January 1738.

19 On Pillai, see Julie Marquet, 'Le rôle des intermédiaires dans l'implantation colonial française. L'exemple de la famille de Tiruvengadam à Pondichéry au XVIIIe siècle', *Encyclo. Revue de l'école doctorale ED 382*, Université Sorbonne Paris Cité (2014), 17–42. On 'dubashes' and 'men of two languages' as 'go-betweens', see Simon Schaffer, James Delbourgo, Kapil Raj, and Lissa Roberts (eds), *The Brokered World: Go-Betweens and Global Intelligence, 1770–1820* (Sagamore Beach, MA: Science History Publications, 2009), pp. 429–40.

20 Ananda Ranga Pillai, *The Diary of Ananda Ranga Pillai*, 12 vols, transl. from the Tamil by order of the Government of Madras, ed. H. Dodwell (Madras: Government Press, 1922), vol. 8, p. 264.

21 For an account of Mughal architecture, see Catherine B. Asher, *The New Cambridge History of India*, Part I, vol. 4: Architecture of Mughal India (Cambridge: Cambridge University Press, 1992).

22 On the French tradition of Orientalism, see Ina Baghdiantz McCabe, *Orientalism in Early Modern France: Eurasian Trade, Exoticism and the Ancien Régime* (Oxford: Berg, 2008).

23 CAOM 26DFC/98, No. 23: Cossigny: Etat a postillé des travaux qui ont été faits, et reparés à Pondichéry depuis le mois de fevrier 1740, jusqu'à la fin d'octobre, et depuis le mois de janvier 1741 jusqu'au commencement d'octobre de la present année, 10 October 1741.

24 CAOM 26DFC/11bis1A, Carte générale des ville, fort et despendances de Pondichéry, [1714].

25 CAOM 26DFC/99, No. 109: Bourcet: Estimation des ouvrages de fortification à faire pour le retablissement de Pondchery, 1765: The precise estimate was 1,437,445 livres, 5 sols, 6 deniers. This included hydraulic infrastructure, too.

26 CAOM 26DFC/99, No. 108: Bourcet: Précis d'un Mémoire relatif à un projet pour le retablissment des fortiffications de Pondichery, 13 October 1765 [not 1762 as dated by hand], fol. 1v: 'Je commence dans ce memoire par expliquer la façon singulière avec la qu'elle on a touhours travaillé dans l'Inde, et les changements qu'il y avoit a faire en proposant de suivre les coutumes pratiqués dans les traveaux du Roy qui sont les seules bonnes et exemptes de friponneries.'

27 Ibid., fol. 2r: 'Je fais voir ensuite la façon de construire ces nouveaux revetements, et combien peu ils couteront en comparaison de cuey que lon construisoit à l'Européene avec les sommes immense sans être meilleures.'

28 CAOM 26DFC/99, No. 104: Bourcet: Mémoire relative à un projet pour le retablissement de Pondichéry, 10 May 1765, fol. 1v: 'C'est a dire qu'il soit dressé un plan général des ouvrages à faire, accompagné d'une estimation exacte, que ces ouvrages souyent donnés par entreprise à des gens capables et en etat de repondre par leurs propres fonds, qu'à la fin de chaque champagne il soit dressé un toisé très circonstancée des ouvrages qui auroient été faits, lequel toisé sera signé par le Gouverneur, les deux premiers Ingenieurs et l'Entrepreneur, et sera ensuitte envoyé à la Compagnie qui le sera examiner très scrupuleusement par des gens du métier.'

29 Ibid., fol. 2r: 'J'ai consulté plusieurs Maitres Maçons Indiens qui se sont assés rapportés entreux, d'ailleurs quant à la maçonnerie, connoissant le prix de la façon de tant de Millers de Briques, de leur transport ainsy quee de la mesure de la chaux et combien un maçon indien pouvoir poser de briques par jour accompagné de ses manoeuvres.'

30 CAOM 26DFC/99, No. 130: Bourcet to the Superior Council of Pondichéry, 19 May 1765: 'Messieurs, les ouvriers des travaux se plaignant continuellement de la modicité de leur paye, qui ne leur permet pas d'entretenir leurs familles, les empêche par consequent de venur s'offrir en quantitié suffisante pour les travaux, aimant mieux server les Anglois, Danois, et Hollandois, qui donne une paye plus forte. J'ay l'honneur de vous proposer, messieurs, de mettre la paye des dits ouvriers sur le meme pied que les Anglois, Danois et Hollandois.' An artisan of the first class received a salary of 1 fanon, 40 caches, which equalled about 15 sols; a female worker received 32 caches, which equalled about 3 sols.

31 CAOM 26DFC/99, No. 130: Abeille to the Superior Council of Pondichéry, 29 June 1765.

32 CAOM 26DFC/99, No. 130: Nicolas Du Laurens: Conditions aux quelles le nommé Ramalingant s'engage pour fourniture des materiaux qui ont rapport aux fortifications de Pondichery, 1 June 1765.

33 CAOM 26DFC/99, No. 227: Bourcet: Mémoires, Toisé, Plan feuille de profils, le tout en un seul cahier, 15 October 1768.

34 CAOM 26DFC/99, No. 132: Abeille to the Company's directors, 12 October 1765, fol. 5r: 'L'intention de la Compagnie a toujours été de former un canal couvert en forme d'egout pour faire degorger au dehors toutes les eaux de la ville, aisni que les immondices, et en même temps concevoir à la salubrité de l'aire, une partie depuis les murs au sud jusques au quartier du Nord ne fomrant qu'un cloac infect, dont on a peine a supporter l'influance dans les temps de la grande chaleur, cet ouvrage très important par luy même devoit etre exécuté suivant l'intention primitive, de

cette façon on communiquerpot avec facilité dans les deux quartiers qu'il sépare, apellé celuy de blancs et celuy des noirs, et si la salubrité de l'air est pour quelque chose dans cequi avoit été ordonné, au canal couvert est à préferer a tout autre, on y laisseroit des regards d'endroit à autre pour en failiter le nétoyement; on luy donneroit la forme indirect ou irrégulière pour l'ecoutement des eaux: Un tel canal ne uniroit à personne, donneroit lieu a des emplacements réguliers pour des maisons qui servoient tant a embelir qu'à augmenter le quartier dit des blancs.'

35 CAOM 26DFC/559A: Louis François Grégoire Lafitte de Brassier: Carte de Pondichéry et ses environs, sur laquelle est marquée l'attaque des Anglais commencée le 8 août 1778 et l'état où était la ville lorsqu'elle a été assigiée, avec tous les travaux qui y ont fait depuis le 5 juillet de la même année, jusqu'au 18 octobre, ca. 1780.

36 CAOM 26DFC/99, No. 131: Etat des Maisons retablies à Pondichéry au 20 août 1765, 15 September 1765. In total 2,929 houses are listed, 126 European houses (65 stone houses with tiled roofs, 61 with thatched roofs), 1,959 Malabar houses (122 stone houses, 1,837 with thatched roofs), 173 small houses for the 'Blancs' (53 with tiled roofs, 120 with thatched roofs), 384 small huts, 44 shops built of stone and covered with tiles, and 243 shops with thatched roofs.

37 CAOM 26DFC/99, No. 226: Abeille to the directors of the Compagnie des Indes, 28 July 1768; CAOM 26DFC/99, No. 227: Abeille: Renvoy détaillé pour server au projet d'un enclose pour la Compagnie à Pondichery, designant les batimens a y construire et leur distribution, 28 July 1768, joint to the previous letter of the same date; 26DFC/224A: Abeille: Projet des bâtiments propres à renfermer les archives, trésors, effets et travaux de la Compagnie, 28 July 1768, originally joint to the previous two documents.

38 Cf. Jean Deloche (ed.), *Le papier terrier de la ville blanche de Pondichéry, 1777* (Pondichéry: Institut français de Pondichéry, 2002).

39 For this chapter of racial discourse in French India from the eighteenth to the twentieth centuries, see Adrian Carton, *Mixed-Race and Modernity in Colonial India: Changing Concepts of Hybridity across Empires* (London/New York: Routledge, 2012), pp. 63–79.

3

Decay and repair: Fort Royal as a perennial construction site on Martinique

The fact that large colonial building structures tended to decay quickly and were not maintained by a private proprietor made the upkeep and repairs very difficult. This chapter will show that early modern building projects on Martinique do not fit in a picture of empire building where a certain idea of planning was executed within a certain cost and time frame. Rather it underlines the importance of repairs and everyday maintenance of different parts of the construction site – a process that sometimes took decades and even centuries.

Cutting costs

The colonial administration in Versailles received quite regular communications about progress of the building works. Fort Royal, which became the new centre for the governor-general of the French Antilles, was a prime concern of the local administration, which kept the secretary of the navy informed by means of several letters, reports (*mémoires*), project designs (*projet*), estimates (*devis*), and opinions (*avis*). These concerned aspects of the economic viability of the projects, such as the high cost of materials, the organization of the workforce, the payment of slaves owned by the island's inhabitants, the defection of slaves and indentured servants (*marrons*), the sustainability of the construction works, and, most frequently discussed in more detail, the financial administration of funds for the building projects.[1]

The work to restore Fort Royal after the battle of 1674 progressed slowly and faced several obstacles. On 22 June 1675, a year after the attack, de Baas wrote in a letter to Colbert that the work on the fortress had ceased because of the death of the foreman who ran the construction site. Since he felt too weak and ill, the governor had not managed to find a replacement. Not one suitable person could be found to continue the work.[2]

Four years later, the difficulties remained, but Jean-Baptiste Patoulet, the royal intendant to the Antilles, presented some solutions to the minister. To resolve the problem of management, the local government employed a contractor (*entrepreneur*) who was paid as the work progressed. But the earthworks needed for the escarpments and the batteries were deemed by Patoulet to be virtually impossible for human labour. For this task, however, he had put to work African slaves that were leased from the plantation owners. From this, in turn, derived a further complication: due to the loss of one twelfth of the labour force on the plantations over a period of eight months, some settlers lacked food for a month or more and others were completely ruined. Patoulet insisted that this was necessary. Furthermore, he would need earth to be transported until May the following year in order to complete the seaward escarpments, the covering of the powder magazine, and the cistern; the African workers had to remain at the construction site for another eight months. His proposal to Colbert for compensating the slave owners of Martinique consisted in an exemption from capitation duty – about 4 livres for each slave on the islands – which would be paid instead by the tax-farmer, the local private financier who obtained the lease for the right to collect taxes.[3]

Patoulet, however, left it open as to whether this was by agreement with the tax-farmers of the Western Domain (*Domaine d'occident*), who were usually, like Jean Oudiette, for instance, influential financiers with strong ties to the capital and the royal administration; Colbert was, as controller-general of finances, one of their closest associates.[4] But the intendant was confident of providing Colbert and the King with a fortification that would make the island one of the best defended places in the world. And all at the cost of 20,928 livres 17 sols 1 denier, as he calculated, a sum that he would beat down to only 14,000 livres, and that François Bellinzani, the head of Colbert's *Conseil du Commerce*, could even pay in the form of merchandise.[5] The expenses that the contractor demanded were carefully negotiated by Patoulet and were conveyed in all detail to Colbert with the rhetorical style of a faithful accountant:

> Since I had the honour, Monseigneur, to let you know that I had reached an agreement with the contractor for the work to fortify Fort Royal of this island that he would in future charge 10 francs less per square toise of masonry than in the past, he has withdrawn from it. I would have brought in someone else to do the work, but as I do not have the authority, he would have met so many obstacles and faced so many difficulties that he was not able to maintain his discount.[6]

The difficulties were obvious: the task was immense. The contractors, therefore, were seeking ways to justify and painted a dramatic picture of the construction site for the local government. This government, usually in disagreement with itself, the intendant arguing in one direction and the governor-general in the other, kept the costs they presented to the administration as moderate as possible in order to appear in a favourable light. And Colbert rarely thought about financial means of support since he wanted the costs for the royal household to be as low as possible, perhaps even nothing at all, as other colonial administrators promised.

In 1688, for example, the intendant Gabriel Dumaitz de Goimpy argued concerning a fortification project in Saint-Christophe that the inhabitants of the island would eventually cover all costs. He even devised a plan to avoid the planters sending only their weakest slaves to the construction site. One would buy new slaves and let them work at the fortress for three years, during which they would be fed and accommodated by the planters. After completion the King would still be owner of these slaves while the whole thing would cost His Majesty nothing at all![7]

In Cayenne, the construction of the fortress progressed very slowly despite having about 500 African slaves at work.[8] The governor François de Lefebvre de La Barre and the engineer in charge, the Sieur de Pasquine, disagreed on how to manage the building project. While the governor complained to the minister that Pasquine would meddle in the financial affairs of the colony, the engineer did propose some alternative measures. First, one should appoint an administrative commissioner (*commissaire-ordonnateur*) and not some clerk of weak character to manage the details of the fund; and secondly, Pasquine wrote, it would be best to allow the American colonists to enter into the slave trade with Africa in order to acquire the resources to pay for the enormous costs of the fortification project.[9] The latter proposal certainly went too far since the lucrative monopoly had been since 1684 in the hands of the *Compagnie de Guinée* and enjoyed the protection of the Marquis de Seignelay, Colbert's son and his successor as secretary of the navy. Despite the benefit for the royal treasury, the administration showed no interest in the engineer's proposal. It did not support the financial situation of the colonies and the monopoly rested in the hands of a few financiers.[10]

So the weight of financing the building projects on the Antilles rested on the community of planters. But it was the workers, African slaves, French indentured servants, and expert artisans, who had to bear the ordeal of the construction itself. The written cost estimates reveal who they were, to what extent they participated, and how they were exploited or could benefit from this work. An estimate from 1771 for the new Fort Bourbon on Morne Garnier gives quite detailed figures: excavation work cost 24 livres a toise, which meant that 7,000 toises would cost 168,000 livres; the excavation of earth came in at 13 livres per toise, resulting in 120,000 livres for 8,000 toises; and the lower lunettes cost 12 livres a toise, adding up to 39,840 livres for all 3,320 toises. Masonry was even more expensive: stone cost 178 livres a toise, that is, 570,150 livres for a total of 3,258 toises; subterraneous walls and portals were to be built using stone at 175 livres a toise and brickwork for the vaults at 130 livres a toise, resulting in 44,500 livres; the terraces were the most expensive part of the fortress: the costs for stone were the same as above, but the costs for the vault's brickwork amounted 250 livres per toise totalling 153,500 livres. All in all, the costs for Fort du Morne-Garnier were estimated at 1,301,590 livres.[11]

The cost projections for the modernization of Fort Saint-Charles and Fort Louis on Guadeloupe during the war were enormous. For the development of Fort Saint-Charles alone one estimate in 1765 came in at 815,505 livres. In 1766, 2,437,596 livres was the estimate for just two bastions above the escarpment leading down to the Galion River, and in August 1769 there was a further estimate of 1,084,278 livres. The bridge over the Galion, built between 1773 and 1780 and still in existence having been only recently restored, had cost about 1,000,000 livres. The cost of refitting Fort Louis was estimated in 1766 at 2,105,866 livres.[12]

Fortification was extremely expensive. If we assume that wages for unskilled labour normally averaged about 300 livres per annum in 1650, rising to 600 livres in 1750, with skilled labour costing roughly 900 livres rising to 1800 livres in the same period, and if we also assume that a high-ranking official would be paid 2,400 to 3,600 livres and a governor up to 12,000 livres, then even the estimates of more than a million livres seem unrealistically low. And this was simply an estimate for the construction of a completely new fortress. It does not take into account the costs for repairs, extensions, and modernizations of the dozen or so fortresses in the French Antilles, which were, of course, of various sizes.

How are these sums to be explained? And was it really possible for the Crown to avoid large expenses? In 1717, the governor-general Antoine d'Arcy de La Varenne wrote to the Navy Council, which took control of the colonies during the regency, stressing the need for the repair and augmentation of the fortifications. On Martinique alone, La Varenne estimated, one would require a fund of 24,000 francs for the fortifications and the maintenance of storehouses, the garrison, and the buildings belonging to His Majesty. When the engineer La Roulais inspected Fort

Royal he found that the residence of the intendant was dilapidated, no longer having doors and windows, and the beams, planks, and the whole framework being in such a bad state that in many places it rained in. The 4,000 francs the Council provided for repairs to this building would not even have covered half of the cost. La Roulais, therefore, asked for another 4,000 francs to be allocated to this fund. 'In this country', La Varenne added laconically, 'very little work is done for a lot of money.'[13]

La Varenne gave a reason for this in a letter he wrote to the Council a month later:

> The Council perhaps finds that the expenses are great and the work not very considerable, but it must grant us the honour of explaining that in this country a mason, a stonemason, an artisan, a carpenter, and other workers of this sort regularly earn 7 and 8 francs per day; and when their Negroes (*negres*) learn something of the occupation from their masters they earn 4 francs and 10 sols per day; thus the toise of masonry two feet and a few inches thick usually costs 33 or 34 livres. Timber, too, costs much more than one could imagine. Therefore, one does very little work for a lot of money on these islands, as I already had the honour of informing [the Council].[14]

Among the new recruits from France there were usually only very few experts in masonry and carpentry, which made them 'quite rare and very expensive on the islands'.[15] We can assume, therefore, that trained African slaves not only carried out the hard labour of excavation, but also the skilled labour of artisan work.

Added to these labour costs, high expenditures for materials were also required. Most importantly, stone, bricks, wood, and quicklime for mortar were quite difficult to acquire. If the building material was not found in the vicinity of the construction site it had to be transported from another part of the island, a different island, or even from France. A report from 1739 by the governor-general Jacques-Charles Bochart de Champigny de Noroy and the intendant César Marie de La Croix informed the minister that the coasts of St. Lucia and of Dominica were almost entirely cleared of forests. To get proper poles and boards one had to go to the remote forests of the interior. It proved impossible for the few Frenchmen on St. Lucia to furnish the larger islands with poles and planks. This posed a difficulty for the development of another large building project in Fort Royal: the construction of a new pier at the Carenage basin as well as a sea wall alongside an area next to the town called Savanne. To hold back the seawater in order to build the pier and the sea wall they had to use wool and small rocks delivered by African boatsmen. That, in turn, led to a less solid construction, which would need much more maintenance than the wooden alternative.[16]

Troubled engineers

Another problem that made the construction of fortresses and other large buildings in the French Antilles more difficult had to do with the tasks and function of the naval engineers that were sent from France in the process of planning, executing, and managing the construction sites. Frequently either the governor or the intendant would urge the administration in France to send qualified engineers. Those experts could be seasoned army veterans, like François Blondel who fought in the Thirty Years' War, studied fortresses all over Europe, and even went on to become Marshal of Camps in 1652.[17] From 1664, however, Colbert established new navy colleges in several port cities in France that trained naval engineers in geography, hydrography, and navigation in order to staff a corps of *ingénieurs de la marine*.[18] Blondel was one of its first members, stationed in the new naval base at Rochefort, where he began building the Corderie Royale in 1666, and leaving after just one year for the Antilles.

But not all the engineers sent to the American colonies were as accomplished as Blondel. Of the Payen brothers, for example, Père Labat found one to be unqualified for the planning of Fort Royal because he was simply an artisan by trade. But the siblings Nicolas, Germain, and Marc were quite a productive team of cartographers. The National Archives keep twenty-six maps that the catalogue attributes to them. Their maps and plans of fortresses and their surroundings exhibit such precision and artistry that they prove the Dominican wrong in holding their expertise in low esteem.[19] Furthermore, the oldest brother, Nicolas Payen, received his royal patent as a fortifications engineer in 1671, and the quality of the maps, profiles, and plans alone bears witness to the geographical and hydrographical expertise the brothers applied in their work.[20]

The Payen brothers fought the same battle against the constant lack of resources as the other officials did. Except the engineers were much closer to the construction site and were confronted with the problems much more directly than others. Their reports about the activities at the construction sites included, therefore, much more detail than those of the island's administration. They, too, asked for more experts, especially masons, to be sent from France and even proposed practical ways and means to reduce the prices of the masonry:

> The price of 60 livres for each square toise of masonry that one had to pay last year appeared to me so excessive that I have endeavoured to

[71]

reduce it by having the material reviewed that comes from so far away to this fortress; but as we have taken a great quantity of stones and pebbles from the escarpment that we are working on the said masonry can be achieved for 30 livres per square toise, provided we be assisted by the fleet which has lately arrived from Rochefort to carry the wood and limestone that must be brought by the Sieur Allouzie and the Sieur Travailler.[21]

Perhaps this provisional solution was the reason for the later deplorable state of the walls that Labat attributed to the doing of Payen. But as the engineers conducted their work at Fort Royal, the pressure to meet the demands of their superiors grew even more. In February 1681 the intendant Patoulet wrote to Colbert that the funds granted for the fortress would suffice for Payen's plans. The price of a square toise of masonry, however, was now pegged at 40 livres, which would be too high since the governor Blénac wanted it capped at 20 livres per square toise.[22] Then, in April, it rained so hard that the masonry on the bastion supposed to protect the fortress to the north-east collapsed and had to be refitted. A profile drawn by Payen in November 1680 shows the problematic geological situation and that two parapets and one rampart were affected. Payen estimated the cost of the repairs alone at 21,208 livres.[23]

On top of these problems the engineer found himself underpaid. Although many people had expressed admiration for his efficient and fast work, he only received 600 livres per month; not enough, he argued, since he had to provide for his two brothers who were also engaged on the project – obviously unofficially and without separate contracts – and had to pay rent of 250 livres per year for a house in the town of Fort Royal.[24]

But in July 1683, Payen appears to have succeeded in refitting at least a part of the collapsed bastion. His plan shows a multicoloured cross section of the structure, No. 25, displaying the different layers of earth: red tuff at the bottom, sand and silt above it, and then the earth that had been heaped on top against the walls, as well as the earth that still had to be moved (see Figures 21 and 22). The engineer wrote to Blénac that six feet of the wall were completed and that thirty soldiers, together with the Africans, had moved earth from the foundation of the flank to fill up the curtain wall of this bastion. Of course, the expense for this achievement turned out to be much higher than estimated, since the soil was continuously moving, and the bastion had to be extended using iron grids in order to stabilize the foundation.[25]

In 1690, the naval officer Louis Ancelin de Gémosac received a letter from Blénac that brought up the problem with the slave workforce. Gémosac, or Gémosat, was also an engineer, who had been in the

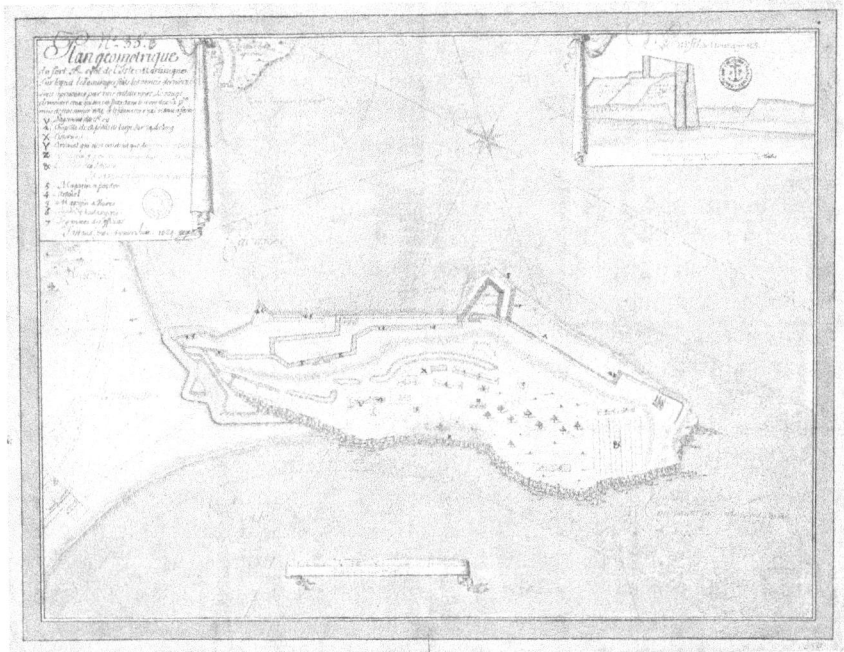

21 Plan géométrique du Fort Royal, 30 June 1685, signed by Payen, attributed to Marc Payen (CAOM 13DFC/35B).

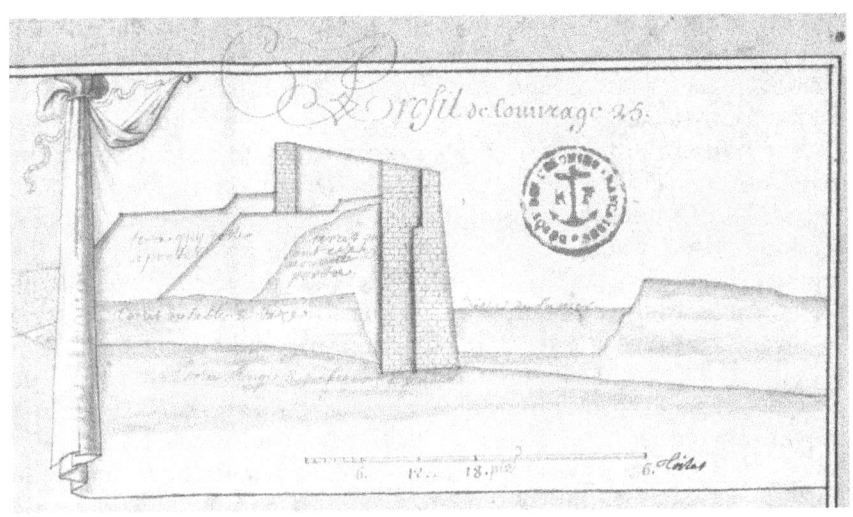

22 Plan géométrique by Payen, Detail.

Antilles since at least 1672, became Lieutenant du roi in 1675, and governor of Grenada later in May 1690.[26] The commanding officer at Fort Royal, Blénac, held him accountable for keeping an accurate record of the day's rates for the slaves leased by the island's inhabitants. Gémosac and Payen both had to provide each other with statements (*états*) of the number of workdays the African slaves spent at the construction site. By this procedure the governor-general hoped to gain better control over the agreement with the planters and to identify those who had furnished only some of their slaves while others remained hidden on the plantations.[27]

The issue of compensating the inhabitants for the use of slaves continued to trouble the chief engineers. About thirty years later, Louis de La Roulais, following Jean-Baptiste Caylus, was in charge of the construction site of Fort Royal, which was still unfinished. The state of the fortress continued to be highly problematic. Caylus even remarked in his letter of 1704 that the flaws in the construction of Fort Royal could only be described in general terms. He would not dare, especially in times like this – it was a time of crisis for France in the War of the Spanish Succession – to confide the imperfections to paper in detail.[28] La Roulais' later attempts to remedy these faults, however, were confronted by the same difficulties faced by his predecessors, except that he now pushed the interests of the large building project with more insistence. He used funds from the navy, generally designated for ship repairs, to pay the garrison etc., to pay for his workers wages, for materials, and for the certificates that exempted the planters who provided him with slaves from taxes.

This brought him up against the navy officials on Martinique, especially the paymaster and the inspector who controlled these funds. Charles Mesnier, *contrôleur de la marine à la Martinique*, was quite adamant in accusing La Roulais of violating the navy ordinance of 1689 that regulated the responsibilities of inspectors and engineers in the colonies very clearly.[29] He wrote several letters first to La Roulais, then to the Navy Council arguing against the irregular assignment of funds, but did not really achieve much. The Council remained silent on the matter. Mesnier, seeing that La Roulais had gained the upper hand, declared that, while the engineer might be better at writing and scheming to advance his own interests, he, for his part, would dedicate himself to following the rules in pursuance of the interests of the King.[30] The governor-general, the Marquis de Feuquières, however, backed Mesnier in pointing out that in contrast to an inspector in Europe he had very clearly defined duties assigned to him. The engineer's pretension to take all matters into his own hands would undermine good order on the islands.[31]

Interestingly, the local administrators saw the trouble with La Roulais as being caused by his unfamiliarity with the customs and traditions of the islands. Mesnier argued for the importance of *anciennité* when it came to administration and regulation, a quality La Roulais did not possess. He, the inspector, had already been thirty-nine years in the service of His Majesty and was motivated purely by his loyalty and zeal to serve the King.[32] But the engineer was a newcomer who applied 'ruses and finesses' and thought to know the country and its inhabitants at first sight. Mesnier deemed this particularly problematic since he thought this would only fuel the already malignant and self-interested character of many of the island's inhabitants.[33] Such behaviour could even lead to 'a second revolt' among soldiers, inhabitants, and slaves, the inspector warned, in allusion to the Gaoulé insurrection in 1717.[34]

Vincent Houel, not related to Charles Houël, the famous governor of Guadeloupe, was sent to Grenada as an engineer in 1719. In 1726, he succeeded La Roulais as officer in charge at Fort Royal, but obtained his patent as captain and chief engineer of the islands only after 1741. The lateness with which Houel's service was rewarded by that promotion, after over twenty years of working on the fortification of the islands, reveals something about the precarious position the engineers had.[35] The budget statements for Fort Royal that he sent to the minister for the navy bear witness to his zeal to balance the costs with the estimates and with the money he was granted by the Crown. In 1729, Houel was able, for example, to cover the costs for all the repairs on Martinique, Guadeloupe, and Marie-Galante from the annual budget of the Western Domain, which was reserved especially for this purpose and amounted to 6,075 livres. Supposedly, the expenses surpassed the budget only by 9 sols and 10 deniers.[36] Including the rebuilding of a wall of the Batterie Royale at the citadel of Fort Royal for which the budget was 9,637 livres, Houel overspent by a margin of only 32 livres 16 sols 6 deniers.[37] The overall picture for the budget was even more favourable. Thanks to a cheaper solution for the repairs to a bastion, the expenses for the year 1729 could be cut by 12,144 livres.[38]

Such calculations may have demonstrated his capacity for accurate accounting and good management, but it also revealed that the costs for the construction and repair of Fort Royal continued to be very high. To transport limestone and firewood, Houel proposed buying a boat for 23,350 livres in order to save the high fees demanded for chartering private vessels. The purchase cost would amortize over time and, even accounting for the cost of maintenance, the navy would save a considerable sum of money, making it a relatively cheap investment.[39]

At first, the engineer's reputation seems to have grown over the years. Houel enjoyed the patronage of governor-general Champigny de

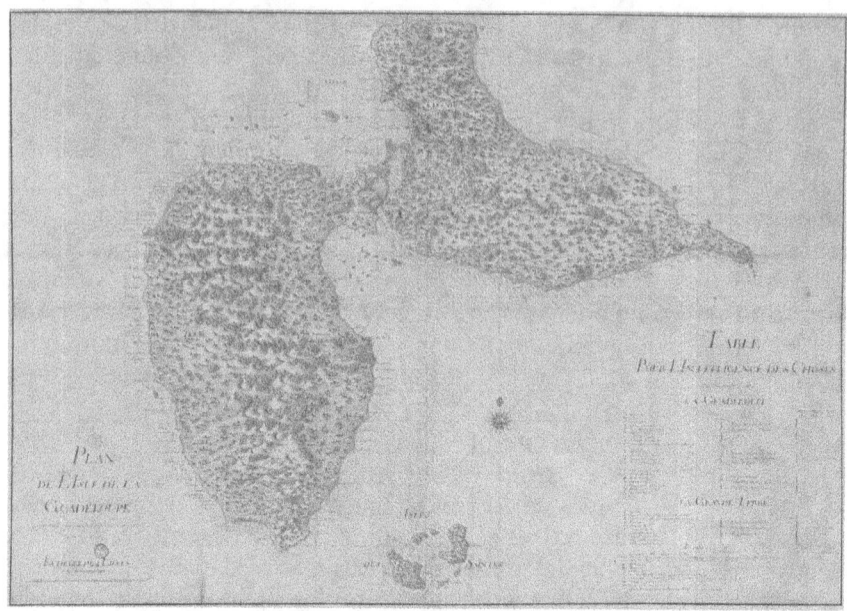

23 Vincent Houel: Plan de l'ilse de la Guadeloupe, wrongly attributed to Jean-Baptiste Houel, 1730 (BNF Paris, Départment Cartes et plans, GE SH 18/PF 155 DIV 2 P 4).

Noroy, who presented the engineer before the minister in the best light. A map Houel had drawn of Guadeloupe, probably of Fort Saint-Charles in Basse-Terre, 'with the utmost accuracy' was proof, Champigny wrote to the minister, that this officer would be worthy of his favour (see Figure 23).

Apparently, someone offered Houel 2,000 ecus for a copy of this map, an offer he seemed to have declined due to his workload and his reluctance to make it public. The governor-general, however, sent the map to the minister, who arranged for a copy to be made in Paris, which was then returned to Martinique, where it was 'kept in great secrecy'.[40]

The engineer's success depended on patronage. With Champigny gone and being succeeded by Charles de Thubières de Caylus as governor-general, he fell into disgrace. Caylus, in fact, did not think much of Houel's qualities as chief engineer on the islands. In 1746 he suspended him, since this 'prétendu ingénieur en chef' had failed to provide him with a general or particular financial account of Martinique and the other islands. The governor complained that this negligence made it impossible for him to find the causes of the large expenditures in the colonies.[41] Houel, not able to exercise his profession anymore, left the islands the same year and did not return. He died in Paris in 1754.

From the perspective of engineers sent from France to the Antilles, large building projects were both an opportunity to advance their military rank, and their place in the administrative hierarchy, and a source for troubles that could eventually lead to their demise. It would be perhaps tedious to answer the question as to whether the methods of financial administration the engineers employed, the ways in which they tracked actual expenditure against the estimated budgets, with surprisingly advantageous results, were actually correct. But the numbers represented only part of the picture of large building projects. The engineers had to be bookkeepers at the same time as doing what they were partly trained for: measuring the landscape, drawing geographical and hydrographical maps, and profiles and plans of buildings and building sites, as well as overseeing the construction of the project itself.

Obviously, this was not always the most rewarding occupation. The engineers had to juggle keeping up with the tasks at hand, giving accounts to their superiors, negotiating with the different interests of the planters and skilled labourers, and coercing indentured servants and slaves to carry out the hard manual labour. But although these experts of large building projects occupied key positions that were very visible to contemporaries and still figure prominently in the sources, they were not the only ones responsible for the success or failure of these projects. Chapter 5 takes a closer look at those experts that, though lacking an official patent as royal engineers, nevertheless contributed considerably to the material construction effort in the colonies.

Notes

1 I have chosen most examples of administrative correspondence from the Archives nationales d'Outre-mer in Aix-Provence (ANOM), where the series of the Dépôt des fortifications des colonies (DFC) and the series of the correspondence Colonies for Martinique (COL C8A and C8B), Guadeloupe (COL C7A and C7B), and Guyane (COL C14) give a good insight into the micromanagement of colonial building projects. The regular correspondence, however, must not be understood as evidence that conclusively proves the success or the failure of colonial control. For the debate among historians about the efficiency or inefficiency of the administrative system for the French colonies, see James E. McClellan and François Regourd, *The Colonial Machine: French Science and Overseas Expansion in the Old Regime* (Turnhout: Brepols, 2010); Banks, *Chasing Empire*; Loïc Charles and Paul Cheney, 'The Colonial Machine Dismantled: Knowledge and Empire in the French Atlantic', *Past & Present* 219 (2013), 127–63; as well as the discussion in Chapter 1.

2 CAOM COL C8A 1: De Baas to Colbert, 22 June 1675, fol. 1v.: 'Le maître qui conduisoit les travaux du Roy au fort Royal s'est noyé par un facheux accident, et depuis ce temps, il m'a esté impossible de m'y trasporter a cause de ma faiblesse, n'y de trouver un seul homme qui me soulage dans un sy pressant besoin, cet ce qui ma contraint de faire cesser les travaux, et de laisser pour un temps les

fortiffications en l'état qu'elles sont durant lequel le Sr. de la Cale dressera les comptes et m'informera de l'état du fonds du Roy qu'il a la main et que je croy fort proche de la fin, La reception à celuy des negres de La Granade a esté retardée par l'entreprise d'Erasmus, et elle ne peut se faire qu'au mois de septembre prochaine, et pour cet effet s'y ennoyeront un des vaisseaux de l'escadre du Roy qui doit venus aux îles, le fonds de quinse negres vendus a la Guadeloupe est deja consommé comme vous les verrez par l'état que je vous en envoyeray.'

3 CAOM COL C8A 2: Patoulet to Colbert, 22 September 1679, fol. 1v.: 'Dans ces travaux [at Fort Royal] il y a deux choses a considerer: les ouvrages de maçonnerie se sont fait à prix fait par un entrepreneur, lequel a esté payé a mesure qui ses ouvrages on advancez, mais que tous les travaux de terre, les grandes escarpes, l'execution des batteries, et l'enterrement des terres qui sont des ouvrages presque incroyables on esté faits par les negres des habitans, dont on a toujours pris la douzieme partie depuis plus de dix huict mois, sans qu'on leur ayt payé aucune chose, cette corvée, Monseigneur, est fort a charge a ces habitans qui ne tirent leur subsitance que par le travail de leurs negres, de sorte qu'on peut dire que la plus grande partie manquent de nourriture un mois de l'année et que d'autres en dont entierement ruinez, par la perte de leurs Negres qui incurent dans ces travaux, cependant ils y seront indispensablement encore employez pour achever les escarpes du coste de la mer, et a porter les terres pour couvrir les magazins à poudre, et la cisterne jusques au mois de may prochain, et encore dix huict mois apres, si vous n'avez la bonté, Monseigneur, d'ordnonner le fonds necessaire pour faire bastir les murs qui doivent occuper toute la hauteur du fort. [...] C'est de cette proposition, je prendray la liberté, Monseigneur, de vous representer que les peubles de pays cy si peu accommodez, qu'ils sont beaucoup de peine de quoy satisfaire au droit de capitation qu'ils toutes les années de quatre livre treize sols neuve denier par teste, et qu'il faux que le fermier pour s'en faire payer les y fasse contraindre.'

4 Cf. Pritchard, *In Search of Empire*, pp. 133–7; Boucher, *France and the American Tropics*, pp. 189–91; for a detailed account, see Daniel Dessert, *Argent, pouvoir et société au Grand Siècle* (Paris: Fayard, 1984), pp. 325–38; L.-P. May, *Histoire économique de la Martinique (1635–1763)*, Thèse pour le doctorat de droit (Paris: Les Presses Modernes, 1930).

5 CAOM COL C8A 2: Patoulet to Colbert, 22 September 1679, fol. 1v: 'Il vous est demandé par le devis que j'ay signé la somme de vingt mil sept cens vingt huict livres dix sept sols un denier pour achever entierement les fortiffication de ce fort, et le rendre une des meilleures places du monde, mais comme j'employeray tous mes soins a bien faire mesnager les fonds qui seront remis, Je vous promets, Monseigneur, de satisfaire entierement a tout cequi est proté par le devis, s'il vous plaist de faire remttre seulement quatorze mille livres a M. Bellinzany qui les fera convertir en marchandises, suivant l'état que je luy en ay envoyé.'

6 CAOM COL C8A 2: Patoulet to Colbert, 9 November 1679, fol. 232r: 'Depuis que j'ay eu, Monseigneur, l'honneur de vous faire sçavoir que j'estois convenu avec l'entrepreneur des ouvrages des fortifications du fort Royal de cette Isle qu'il donneroit la toise cube de maçonnerie a dix francs moins à l'advenir que par le passé, il s'est retracté. J'aurois bien fait faire la chose par une autre, mais comme je n'en suis pas le maistre, il y auroit tant trouvé d'obstacle, et on luy auroit fait tant de dificultez qu'il n'auroit pas pû soustenir son rabais.'

7 CAOM COL C8A 5: Dumaitz de Goimpy to Seignelay, 28 July 1688, fol. 104: 'Les habitans de la basseterre se sont assemblez de nouveau et ont accordé la demande qui leur a esté faits de fournir la quantité de sucre qui sera necessaire pour la construction de ce fort. Ces habitans ont peu de Negres et demandent qu'il plaise de fournir les 70 qu'ils ont promis et qu'ils les payeront a 13 livres et 10 sous par mois comme au Fort Royal. M. Dumaitz trouve cette proposition avantageuse en ce que les habitans n'envoyeroient au travail que les plus faibles de leurs Negres, et qu'il y en manqueroit tousjours; Son avis seroit donc d'achepter ces 70 noirs qui seroient payez par leur travail en trois ans pendant lesquels ils seront nourris et etnretenus et appartiendroient ensuitte au Roy sans qu'il en cousta si rien a sa Majesté.'

8 CAOM COL C8A 5: La Barre to Seignelay, 12 February 1690, fol. 111v.
9 CAOM COL C8A 5: Pasquine to Seignelay, 7 February 1690, fol. 112v: 'Il seroit bien necessaire qu'il y eust sur le lieu un commissaire ordonnateur qui pust entrer dans le detail de touttes choses et non un escrivain qui n'a pas assez de caractère. [...] Il a desja fait sçavoir que la despense de toutte la fortification reviendra y compris le fer et la charpente a environ 90 milles livres dont si le Roy veut payer les tiers montant a 30 milles livres on en employera 18 milles livres a acheter 300 Negres qu'on distribuera aux particuliers de la colonies a proportion de ceux qu'ils auront fournir et les douze autres mil livres reseront sur les lieux. Il n'en cousteroi rien a Sa Majesté si on permettoit a des particuliers le negoce de la coste de Guinée.'
10 In this case it was a tax-farmer who profited. Claude Cébéret du Boullay, the principal shareholder of the Guinea Company, held most shares of the Western Domain, the old lease of Jean Oudiette, which was then integrated in the General Tax Farm (*Cinq Grosses Fermes*); see Kenneth Banks, 'Financiers, Factors, and French Proprietary Companies in West Africa, 1664–1713', in Roper and Van Ruymbeke (eds), *Constructing Early Modern Empires*, pp. 95–6.
11 CAOM 13DFC/49, No. 322: Estimation des ouvrages proposés pour parvenir à l'entière éxécution du Fort Bourbon, d'après le projet du Ministre du 22 septembre 1770, 1771.
12 See CAOM 08DFC/27, No. 204, 238, and 271.
13 CAOM COL C8B 22: La Varenne to the Conseil de la Marine, 21 February 1717, fol. 39r: 'A son retour Mr. l'Intendant et moy aurons l'honneur d'envoyer au Conseil les projets et devis des reparations a faire, et de l'Augmentation des ouvrages que ledit Sr. de la Roulaye jugera absolument necessaire pour donne une espece de seureté a la Colonie. Si nous jugeons des travaux des autres isles par ceux qu'il faut faire a la Martinique, nous devons avertir le Conseil qu'il faudra au moins tous les ans un fond de quatre vingt mille francs pour les fortifications et entretien des magasins, corps de garde et maisons apartenant à Sa Majesté. Le Sr. de la Roulaye a visité celle dans laquelle est logé Mr. l'Intendant, qui est délabrée de toutes parts n'y ayant plus ny portes ny fenestres qui puissent servir, non plus ue les poutres les planches la charpente et les couvertures qui sont en si mauvais estat qu'il y pleut en quantité d'endroits. J'ay l'honneur de représenter au Conseil que les quatre mille francs qu'il a eu la bonté d'accorder a Mr. l'Intendant pour les reparations de ladite maison, ne fourniront pas a la moitié de la dépense necessaire pour la mettre en estat. Mr. l'Intendant seroit tres obligé au Conseil s'il vouloit bien avoir pour agreable d'ordonner encore une somme de quatre mille francs dans les premiers fonds qui seront remis pour l'Entretien des fortifications. On fait en ce pays très peu d'ouvrage pour beaucoup d'argent.'
14 CAOM COL C8B 22: La Varenne to the Conseil de la Marine, 21 February 1717, fol. 96: 'Le Conseil trouvera peutestre que la dépense sera grande et les travaux pas beaucoup considérables, mais il nous permettra d'avoir l'honneur de luy représenter qu'en ce pays icy un Maçon, un tailleur de pierre, un charpentier, un menuisier, et autres ouvriers de cette espece gagnent regulierement par jour sept et huit francs, et leurs negres quand ils commencent a sçavoir un peu du mestier de leur maistre gagnent quatre francs et cent sols par jour, ce qui fait que la touse courante de maçonnerie de deux pieds quelques pouces depaisseur couste trente trois et trente quatre livres. Le bois de charpente couste aussy beaucoup plus que l'on ne peut se l'imaginer, ainsy comme nous avons desja eu l'honneur de le mander on fait aux isles fort peu d'ouvrages pour beaucoup d'argent.'
15 Ibid.: fol. 98: 'Lors qu'il plaira au Conseil envoyer des recreües en ce pay icy, Mr. l'Intendant et moy le suplions tres humblement et avec instance de vouloir donner des ordres précis pour que dans les dites recreües il y ait beaucoup de maçons, de tailleurs de pierre, de charpentiers, et de menusiers, qui sont tous fort rares et tres chers dans les isles de l'Amerique, ou toutes sortes d'autres ouvriers seront d'une grande utilité. Il seroit a souhaiter en mesme temps que chacun des dits ouvriers eust des outils.'

[79]

16 CAOM COL C8A 50: Champigny de Noroy/La Croix to Maurepas, 15 April 1739, fol. 37: 'Suivant les reponses que nous avons eu des isles de Sainte Lucie, et de a Dominique, les Bords de la mer de cette dernière isle sont presque entièrement defrichés et habitués par des François, en sorte que les Bois sont fort reculés, et qu'on ne pourroit en tirer les pieux et madriers que nous avions demandé sans des depenses considerables supérieurs aux forets des françois qui y sont établis. L'isle de Sainte Lucie n'est pas dans le même cas, mais ceux de nos habitans qui y ont des chantiers ne seroient pas en etat de nous fournier les pieux et madriers dont nous avons besoin, tant pour le quay que vous avés ordonné de faire au basin de carenage du fort Royal à le longe du Morne des Capucins, et pour la digue à pratiquer le longe de la Savanne de ce même foret, à fin d'empecher la mer de la ruiner comme elle fait, que pour les ouvrages mentionnés dans nôtre lettre chiffrée du 4. Janvier dernier, en sorte que pour diminuer l'objet de ces pieux, et madriers, nous projettons de faire executer le quay le long du Morne des Capucins, et la Digue de la Savanne avec des laines et des pierres perdues, vont l'ouvrage ne sera pas moins solide, et sera cependant d'un entretien beaucoup moindre que celuy des qua, et digue des bois projettés, et occasionnera, cependant moins de deboursés, puis qu'il s'executera par les Negres de Corvée, que ce seront les bateaux de corvee qui fourniront les Roches, et que les Bois nécesaire pour les caisses seront d'un petit objet. Nous auront incessament l'honneur de vous en envoyer les état éstimatifs pour vous mettre enétat de déceder sur la préference a donner entre l'ancien projet et le nouveau.'

17 In France they were called *ingénieurs du roy* since about 1550 and were responsible for fortification in the first place. During the reign of Louis XIII their number increased, but formed an organized corps only after 1650. See David Buisseret, 'French Cartography: The *ingénieurs du roi*, 1500–1600', in David Woodward (ed.), *The History of Cartography* (Chicago, IL: University of Chicago Press, 2007), vol. 3, pp. 1504–21; David Buisseret, *Ingénieurs et fortifications avant Vauban. L'organisation d'un service royal aux XVIe–XVIIe siècles* (Paris: Éditions du C.T.H.S., 2002); Anne Blanchard, *Les ingénieurs du 'roy' de Louis XIV à Louis XVI. Étude du corps des fortifications* (Montpellier: Déhan, 1979), pp. 33–70.

18 Cf. Michael S. Mahoney, 'Organizing Expertise: Engineering and Public Works under Jean-Baptiste Colbert, 1662–83', *Osiris* 25, Special Issue: Expertise: Practical Knowledge and the Early Modern State, ed. Eric H. Ash (2010), 149–70; Charles Bourel de La Roncière, 'Origines du service hydrographique de la marine', *Bulletin de la Section de Géographie* 31 (1916), 6–28. The superintendance for fortification was divided under Louis XIV between Colbert (Navy Department) and Le Tellier (and later on Louvois, War Department) and was combined again within an independent Department for Fortifiaction after the death of Louvois in 1691, which was to become the Corps Royal du Génie in 1776 (see Blanchard, *Les ingénieurs*, pp. 60–1 and 71–114).

19 The plans are not always to be attributed to a specific sibling. It is most probable that Nicolas Payen signed his maps with 'Payen' until 1688, the year of his death. Most notable are:
  – a plan of Fort de la Madeleine and the surrounding area in Guadeloupe (CAOM 08DFC/14B: Plan géométrique du fort de la Magdelaine avec ses environs au quartier du Baillif de l'Isle de la Guadeloupe, 31 May 1682, signed Payen, attributed to Marc Payen);
  – two maps of the Cul du Sac that seperates the two parts of Guadeloupe and show the site of the later Fort Louis (CAOM 08DFC/13B and 16B: Plan géométrique du petit cul du sacq qui sépare la grande terre de la Guadeloupe, 31 May 1682 and 26 June 1686, both signed Payen, attributed to Germain Payen);
  – a plan of the Bourg de la Basse-Terre on Guadeloupe with the castle of Charles Houël (CAOM 08DFC/15B: Plan géométrique du bourg de la Basse-Terre de la Guadeloupe, 31 May 1682, signed Payen, attributed to Germain Payen);
  – a slightly different one dated four years later (CAOM 08DFC/17A: Plan géométrique [...], 26 June 1686, signed Payen, attributed to Germain Payen);

- a very exact plan of Fort Saint-Pierre (CAOM 13DFC/48B: Plan géométrique du fort Saint-Pierre, 28 July 1687, signed Payen, attributed to Marc Payen);
- a map of the whole Bourg de Saint-Pierre (CAOM 13DFC/41A: Plan géométrique du bourg et fort de Saint-Pierre, 30 September 1685, signed Payen, attributed to Marc Payen);
- two similar plans for a small fortress projected for the Bourg and Cul du Petit Goave on Saint-Domingue (CAOM 15DFC/699B and 700B: Plan géométrique du bourg et cul du Petit Goave, both dated 6 April 1688, the first signed by Marc Payen);
- two similar plans of a castle for the Bourg du Port Paix also on Saint-Domingue (CAOM 15DFC/434B and 435B: Plan géométrique du chasteau et bourg du Port-de-Paix, both dated 16 April 1688, the first signed by Marc Payen); and
- several plans of Fort Royal and its surrounding area from the years 1681 to 1691.

20 One of the younger Payen brothers received his patent (*brevet*) as engineer for Saint-Domingue in 1694 (CAOM COL B 18, fol. 11: Brevet d'ingénieur à Saint-Domingue, 1 January 1694).

21 CAOM COL C8A 2: [Nicolas] Payen: Projets de travaux pour le Fort Royal, 30 November 1680, fol. 416r: 'Le prix de chaque toise cube de massonnerie que l'on payoit l'année derniere à soixante livres, m'a paru sy excesif que je me suis appliqué à le diminuer ayant examiné les matteriaux quy sont fort esloignés dudit fort, mais comme nous tirerons quantité de pierre et caillous de l'escarpement des ouvrage que je propose, ladit massonnerie se pourra faire a trente livres la toise cube pourvue que nous soyons aydes de la flutte qui est arrivée depuis peu de Rochefort pour voiturer les bois et pierre a chaux qu'il faut apporter de Sr. Allouzie et de Sr. Travailler.'

22 CAOM COL C8A 3: Patoulet to Colbert, 10 February 1681, fol. 63v: 'A l'arrivée de Mr. Payen je le pray de faire un nouveau loizé de tous les ouvrages, qui ont été faits aux fortification du fort Royal, par l'Entrepreneur, qui en estoit chargé. Et comme il s'en trouvé quelque difference sur celuy de l'année passée au desavantage du Roy, j'ay dressé un nouveau compte de la despense qui s'est faite pour lesdites ouvrages. Vous verrez, Monseigeneur, s'il vous plaist de vous le faire representer et du jetter les yeux sur l'Arreste dudites compte que toutes le despenses faites depuis l'arrivée de Mr. le comte de Blénac jusqu'à cette heure payées, il reste a somme de 4222 livres et laquelle sera joint a celle qui proviendra de monte du noveau fond des 20000 livres que le Roy a fait. Ce deux sommes seront plus que suffisants pour achever les ouvrages projetez par led Sr. Payen. Je n'ay pas fait de marché pour ces ouvrages parce qu'il ne ce seroit pas trouvé d'entrepreneur a mains de 40 livres la toize cube de maçonnerie, au lieu que par les secours que Mr. le comte de Blénac donne elle n'en considra au plus que vingt.'

23 CAOM COL C8A 3: Payen to Colbert, 9 December 1681, fol. 118: 'Conformement a vos ordres du 26e juin dernier je me suis resenté a Mrs. Le Comte de Blenac et Patoulet pour apprendre d'eux les intentions du Roy sur le sujet des travaux du fort Royal de cette ise, Ils m'ont ordonné de travailler incessament à revestir de massonerie les escparpes faittes les années dernieres d'autant qu'elle sont en partie renversée n'estant que de tufe rouge et sablonneux quy s'est detrempé par les pluyes quy sont surgenues depuis le mois d'avril dernier qvec tant d'abondance qu'elles ont deschaussé le fondement des vieux murs quy sont audessus desittes escarpes, j'en ay donné avis à Vostre Grandeur dans la lettre que je me suis donné l'honneur de luy escrire en datte du permier october dernier, Elle trouvera dans cellecy un estat que j'ay fait par estimation de ce que se montera le vevestement dedittes escapes quy est de 21208 livres 6s. 8 d.'

24 Ibid., fol. 119v: 'Plusieurs personnes sont surprises de ce que nous avons fait tant d'ouvrage en sy peu de temps, ou nous n'avons despensé jusqu'à present que quinze mille livres. Je continue a supplier vostre Grandeur d'augmenter ledite appointement qu'elle me donne, m'estant impossible de subsister en ses cartier à moins de 600 livres par mois avec mes deux freres qui sont actuellement appliqué à la conduitte des ouvrages.'

[81]

25 CAOM COL C8A 3: Payen to Blénac, 15 July 1683, fol. 119v: 'Trente soldats, et les Negres travaillent a tirer les terres du fondement du flanc dud. demy bastion, et ensuitte a aproffondir celuy de la courtine, qui le joint au redan, qui luy doit servis de flanc et a la batterie Royalle, cette depence sera plus forts que l'estimation qui en a esté faitte, a cause que les terres sont mouvantes et remplyes de sourdis, ce qui nous obligera à les approffondir, et eslargir beaucoup plus qu'il n'avoit esté projetté, et mesme ay poser du grillage et pyllotis, pour en affermir le fondement.'

26 On Gémosac and his important voyage and reconnaissance mission to West Africa, see Gérard Chouin, *Colbert et la Guinée. Le voyage en Guinée de Louis de Hally et Louis Ancelin de Gémozac (1670–1671)* (Saint-Maur-des-Fossés: Sepia, 2011).

27 CAOM COL C8A 3: Blénac to Gémosac, 30 January 1690, fol. 209r: 'Il ne se peut faire qu'll n'y aye de la conformité entre ses deux estats, puis que le sr. de Gemosat ne s'est conduis dans les commandements qu'il a fait aux capitaine de milice pour avoir les negres des habitants que suivant ce luy que les Sr. Payen luy a envoyé ainsy qu'il se peut justiffier par l'ordre de Monsieur le comte de Blenac du 15e janvier ou il demande que 2159 jornées de reste de 62650 que les habittans s'estoient obligez de fournir au mois de juin 1689. A l'esgard de ses Negres secellez il n'y a en qu'une partye fournie par quelques habittants Monsieur de Blenac n'ayant pas voulu donner ses ordres pour faire venir les restant des Negres caché, quoy que le Sr. de Gemosat luy aye representé qu'il ne pouvoit convaindre les habittans a les fournir sans un ordre expres de luy d'autant que c'estoient des privilegiez.' Gémosac wrote in the margin of this letter: 'Je donneray au Sr. Payen l'estat des journées des Negres qui on este forniers au fort Royal affin qu'il le confront avec le sien, et je verray en quoy ils disconviendront Mr. Payen m'a dit avoir receu de vous l'estat des Negres scellez, il vous en rendra raison a la fin de ce mois.'

28 CAOM COL C8B 2, No. 72: Jean-Baptise Caylus: Mémoire touchant la redoute du Morne des Capucins, July 1794: 'J'ay ecrit en general les deffauts de la construction du fort Royal, mais je n'ay jamais ozé, et oze encore moins a present, confier au papier le detail de ces imperfections.'

29 CAOM COL C8A 21: Charles Mesnier to La Roulais, 7 June 1717, where he informs the engineer about the article of the *Ordonnance de la Marine* that one would follow in this country. La Roulais remarked defiantly in the margins that Mesnier had confused the navy ordinance with the one for fortification; the former did not say anything about fortification.

30 CAOM COL C8A 21: Mesnier to the Navy Council, 19 June 1717, fol. 310: 'Il m'est revenue que ledit Sr. de la Roulaye avoit dit hautement en presence des officiers de la garnison du fort Royal, qu'il avoit écrit contre moy au Conseil et à Monsieur le Directeur des fortifications. A cet égard je supplie tres respectueusement le Conseil de vouloir bien faire attention que mon employ me donnant lieu d'entrer en discution pour les interêts du Roy, avec plusieurs personnes presque tous de differents genies, il est comme impossible que je n'en mécontente pas quelques uns qui sont grands ecrivains, à moins que je ne me relâche sur les ordonnances du Roy, et que je veuille permêttre des abus contre ses interêts et l'ordre de son service. J'aurois lieu d'etre fort content de M. de la Roulais s'il a observé la verité dans ce qu'il a écrit, étant persuadé qu'il n'auroit pû mieux s'y prendre pour faire ma cour.' The Council reacted only insofar that one should take the practice in Canada and the Île Royale as an example for that in the Antilles ('Regler leurs fonctions comme ils sont été reglés à l'isle Royale', someone wrote in the margins of the same letter).

31 CAOM COL C8A 24: François de Pas de Mazencourt de Feuquières to the Navy Council, 18 December 1718: 'Voilà en general quelles on esté les fonctions dudit Inspecteur que Mr. de la Roulais luy veut absolument interdire, et veut le reduire aux simples fonctions de piqueur, en suivant uniquement les ouvriers et journaliers et rendant exactement compte à luy seul. Mr. de la Roulais a des prétentions encore bien plus particulieres, qui nous paroissent toutes nouvelles et tout a fait contraires au bien au bon ordre qui s'observe dans nos ports de France, et aux odronances de la marine que nous devons exactement observer icy comme ailleurs.'

32 CAOM COL C8A 25, Mesnier to the Navy Council, 13 March 1718, fol. 57.

33 CAOM COL C8A 25, Mesnier to the Navy Council, 20 March 1718, fol. 62: 'Je suplie très humblement le Conseil d'estre persuadé que si la disposition de touttes ces choses estoit entre les mains d'un Ingenieur en ces pays cy, les consommations iroient grand train particulierement quand cet Ingenieur n'y est pas ancien, les ruses et les finesses de ceux qui l'habitent sont au dessus de ce qu'on en peut dire. J'oserois asseurer à cet égard le Conseil qu'une des chose qui a le plus fait de mal est que les nouveaux venus croyent du premier coup d'œil connoître et le pays et ses habitants, veulent consequemment à cette idée travailler sans prendre de Conseil que de leur teste et ne manquent jamais de gaster tout ce dont ils se meslent et toujours aux depens des interets du Roy et fort souvent à celuy de tout le monde. C'est certainement un mal, car il n'y apoint de pays au monde ou les precautions et la prudence soient si necessaire que dans celuy cy, tout le monde y étant egallement couvert et maline, et extraordinairement attaché à ses intérets.'

34 Ibid.: Mesnier to the Navy Council, 22 March 1718, fol. 72f. For the revolt of the Petit Blanc on Martinique, see Banks, *Chasing Empire*, pp. 34 and 138; Jacques Petitjean-Roget, *Le Gaoulé: La révolte de la Martinique, 1717* (Fort-de-France: Société de l'histoire de la Martinique, 1966).

35 He regularly asked for the commission of chief engineer at least since 1729, but the minister, although continuously urged by the governor-general Champigny himself, ignored Houel's pleas.

36 CAOM COL C8A 41, fol. 311: Vincent Houel: Etat des ouvrages ordonnés par le roi pour les fortifications des îles de la Martinique, de la Guadeloupe et de Marie Galante pendant l'année 1729, 1 November 1730.

37 Ibid., fol. 451: Houel to Maurepas, 10 November 1730.

38 Ibid., fol. 299: Houel to Maurepas, 1 November 1730.

39 Ibid., fol. 304: Houel to Maurepas, 1 December 1730, here fol. 306–7r: 'Je n'ay pu me dispenser d'y mettre l'achat et l'entretien d'un bateau pour le transport d'une paprtie de la roche a chaux et du bois a bruler parceque les bateaux qui naviguent au tour de l'isle ne peuvent sufire pour aporter toute la quantité nécessaire a cause qu'ils sont en petit nombre de sorte qu'ils n'en pourroient pas asse fournir pour l'employ des fonds de chaque année par la construction du fort, ce bateau augmente l'estimation de 23350 livres vous verrez Monseigneur par l'achat qu'il est tres petit parce que je compte que vous aurez s'il vous plait, agreable d'ordonner qu'on tienne la main a ceque tous les bateaux et chaloupe ded. Vaisseaux qui naviguent a la Guadeloupe aportent chacun un voyage de roches a chaux comme l'usage a eté de tout temps, et comme il se pratique a la Martinique pour le transport des roches pour led. fortification en prenant par les Messieurs des Bateaux et patrons de canots un certificat de l'ingenieur qu'ils sont enregistre au bureau des classes pour les decharge des proprietaires, a quoy Mrs. les Generaux et l'intendant tiennent la maine severement dans cette isle, car si l'on etait obligé d'acheter a la Guadeloupe la chaux tout faite ou payer la voiture des rochers a chaux et du bois a bruler ce qui reviendroit a peuprès a la meme chose, il en couterout au Roi suivant le toisé 75640 barils de chaux a 3 livres le Baril qui monteroient a la somme de 223930 livres la quelle jointe a celle de 216385 livres 3. 9. Cela feroit en tout 440304 livres 3. 9. Cette grande difference, Monseigneur, ma fait prendre le party d'employe dans cette estimation un bateau avec tous son entretien pendant 5 années à commencer du premier janvier 1732 ayant suffisament de la chaux brulée et a bruler.'

40 CAOM COL C8A 42, fol. 127: Champigny de Noroy to Maurepas, 8 October 1731, here fol. 128r: 'Vous remarquera Monseigneur par le Plan que vous recevez de la Guadeloupe que M. Houel vient de faire avec la derniere exactitude, combien cet officier s'efforce à se rendre digne des graces que vous avez envie de luy procurer; je sçay qu'on luy a offert 2000 ecus de cette carte, mais comme il ne convient point qu'elle soit rendue publique, et que les grandes occupations qu'il a d'ailleurs ne luy permettent pas non plus d'en faire une pareille pour rester dans le gouvernment, je vous suplie, Monseigneur, d'en vouloir bien faire tirer une copie aussi bien que de celle de la Martinique que vous aurez la bonté de nous envoyer et que je conserveray tres secretement.' Both maps of Guadeloupe, the original and the copy, as well as the

map of Martinique mentioned here are now in the BNF Paris, Département Cartes et plans: (1) Plan de l'ilse de la Guadeloupe, wrongly attributed to Jean-Baptiste Houel, 1730, GE SH 18/PF 155 DIV 2 P 4; (2) Plan de l'isle de la Guadeloupe copié sur celui envoyé par Mr. Houel, attributed to François-Pierre Le Moyne, 1730, GE SH 18/PF 155 DIV P 5 D; (3) Plan de l'isle de la Martinique, wrongly attributed to Jean-Baptiste Houel, 1729, GE SH 18/PF 156 DIV 2.

41 CAOM COL C8A 57, fol. 41: Charles de Thubières de Caylus to Maurepas, 24 February 1746, here fol. 42r: 'J'ay oublié de vous parler dans la lettre ou je vous ay rendu compte de l'interdiction de M. Houel, d'un des griefs que j'avois contre ce pretendu ingenieur en chef, qui n'a pas été en état a mon arrivée de me remettre aucun plan general ny particulier de la Martinique ny d'aucune des autres isles de mon gouvernement. Cequi, je ne doute pas, vous paroistra aussy singuler qu'a moy. Cela est d'autant plus facheux que je seray obligé d'en faire lever cequi causera de la depense et nous ne sommes pas dans un temps et dans une situation a en faire, a peine pour vous nous suffire au plus necessaire.'

4

Mixed society and African 'Rococo': 'French' style in Saint-Louis and on Gorée Island

The emotional effect an empire had on its subjects depended to a large degree on the style and material of the large colonial buildings. The mixed society of the Senegambia region presents an example where the confinement of colonial spaces served the French and local collaborators to create an emotional community. This chapter explores how style and ornaments added to the function of so-called affective buildings that were built both under European and African influence.

French building activity in early modern Senegambia seems to have been rather modest.[1] Compared even to the Antilles, the fortresses were not only smaller, but also less obviously 'French' in terms of their architectural language. In the Senegambia region, different European and African styles overlapped and combined to form a distinctive Creole style that perhaps contained more indigenous elements than was the case in other places.[2] 'Portuguese' style was already established when French traders founded their posts on the Senegal River and Cape Verde in the early seventeenth century.[3] For some time the architecture of their fortresses, government mansions, storehouses, etc. remained simple. But within just a short period of time, the French colonial entrepreneurs of the Senegal Company were able to leave their mark on the building history of the region, and on the island of Gorée they were eventually to be successful in developing a distinct French colonial style.[4]

In 1664, the settlement of Saint-Louis, founded by merchants from Rouen on the island of N'dar at the mouth of the Senegal River, became the property of the West India Company.[5] This establishment served as a trading post that the private entrepreneurs from Normandy had maintained for many decades, perhaps even for as long as two centuries, as the young navy lieutenant Jean-Baptiste Ducasse claimed to have heard local people say.[6] Ducasse arrived there on his second voyage to

Africa in 1687 and seemed quite aware of the fact that any enterprise, be it for trade or the construction of larger buildings, required the assistance of and co-operation with the Africans of this land. In his *Relation* from this journey he points out the necessity of local expertise to navigate the Senegal River and access the market of mostly commodities like gum arabic, leather, ivory, ostrich feathers, gold, and, at this time, only few slaves.[7]

The island of Gorée off the coast of Cape Verde had been in Dutch possession until it was taken in 1677 by Jean d'Estrées, who afterwards demolished its two fortresses. Gorée was an important entrepôt and stronghold for the otherwise unfortified Lusoafrican trading villages of Rufisque, Portudal, Joal, and the posts on the Gambia River. Ducasse proposed to rebuild the houses and defences of the island, since it was in a state where it 'could not defend itself against a sloop'.[8] In the last sentence of the report he indicates that the expense of such a fortification would be very low since stone and chalk would be found on site.[9]

But it was not part of Ducasses's responsibilities to further pursue the building projects on this coast, and fortification works took the French longer than at other places. In 1693, when English forces took possession of Saint-Louis and Gorée for a short period of time, the director, Michel Jajolet de La Courbe, estimated the defence works at both places as minimal. At the bar of the estuary of the Senegal River there was only one battery defending the fort against smaller vessels; in the settlement itself there were only thirty men and no fortification whatsoever to defend the place.

Furthermore, if they retreated to Gorée they were no safer, since the fortifications were not finished and could not resist an enemy attack for very long.[10] During the operation to retake these places many French soldiers fell ill, and La Courbe as well as Ducasse employed free and enslaved Africans to operate cannons and install fascines (bundles of sticks or pipes used to fill in marshy ground) for them.[11] The Brak, a local ruler near Saint-Louis, also provided laptots (African colonial troops) to navigate the difficult waters, which largely contributed to French success in expelling the English from Senegal.[12]

African assistance was also essential for the building of a fortress after the recapture of Saint-Louis. Its construction was conducted during the tenure of Louis Moreau de Chambonneau as director of the settlement. He sketched the fort on several plans from two perspectives and made his depiction of the plain structure look as imposing as possible (see Figures 24 and 25). The familiar colonial habit of placing the representational emblems above the entrance gate is perhaps the most ornamental feature of this edifice. But the King's coat of arms is not in the centre, but to the left, while a figure of the Madonna with Child

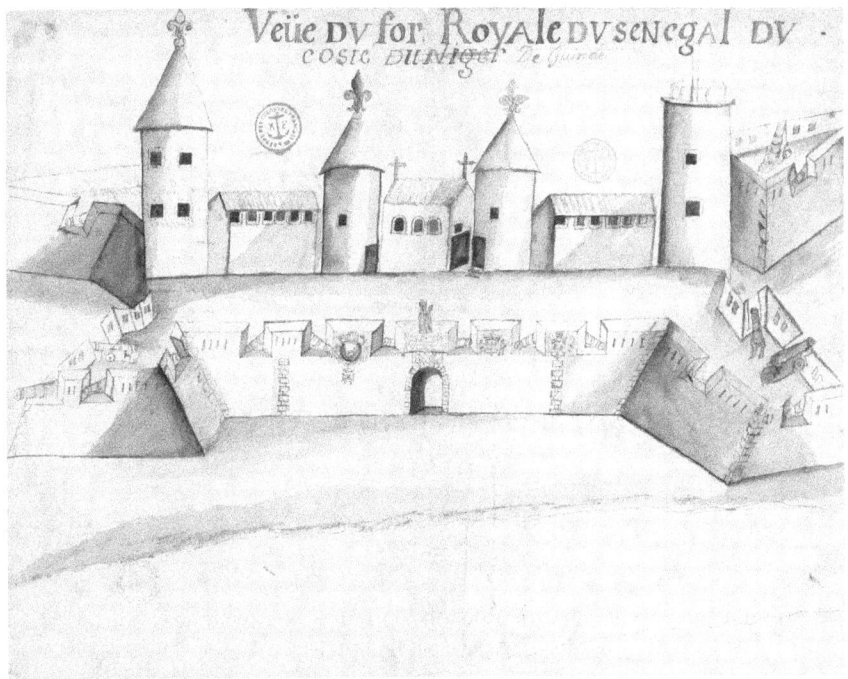

24 [Louis Moreau de Chambonneau]: Vue du Fort Royale du Senegal du
Coste de Guinée, 1694 (CAOM 19DFC/10C).

occupies the most prominent position. To the right there is a rare
depiction of the emblem of the Senegal Company and the coat of arms
of Chambonneau himself. The entire complex comprised four towers
of medium height, a simple chapel, two small buildings, and five
bastions.

Chambonneau himself did not focus his efforts on the development
of Saint-Louis Island. Instead, he travelled upriver to form trade agree-
ments with the local rulers of the interior like the Damel, Siratik, or
the king of Galam.[13] When La Courbe came to relieve Chambonneau
from his post in 1685 (both rotated the command several times) he
saw that the fortress was in a state of disarray. It was open to all sides,
with Africans and whites intermingling, European employees forming
households together with Africans, residing in the straw huts outside
the compound rather than inside the fortification, and with the Company's agents dressing casually in only shorts and shirts.[14] Later on, in
1694, La Courbe – one of the more ambitious directors of the colony
– sought to improve the defence of the river and even planned a second
fortress on an adjacent islet.[15]

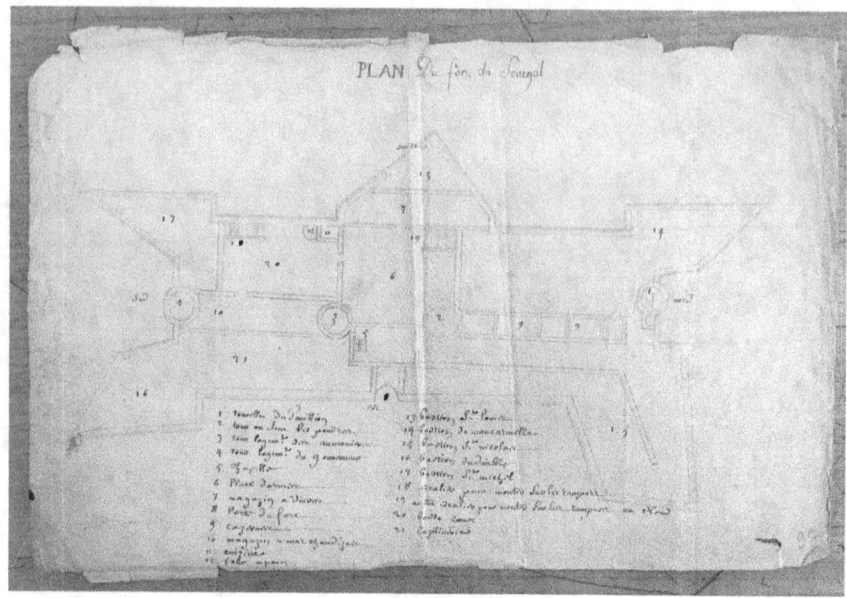

25 [Louis Moreau de Chambonneau]: Plan du Fort Royale du Senegal, 1694
(CAOM 19DFC/9C).

Instead of this project, a different proposal for Saint-Louis by François Froger, an engineer-hydrographer known for his accounts of voyages to Africa, Brazil, and the Antilles, and for being among the first French travellers in China,[16] was brought forward. In 1704, Froger was commander of the vessel *L'Amazone* and sailed for the coast of Senegambia. Here, he drew up two plans for Saint-Louis: one for the old fortress and another for a new one (see Figure 26). Froger's view of the old fortress shows few differences to that of Chambonneau, but it tells us more about the surrounding area. Outside the north-west bastion of the irregularly shaped fortress was a geometrical garden, and to the south several small houses were scattered in no apparent order. A closer look, however, reveals a typical African system of enclosure of groups of houses called *tata*, which was, as we know from archeological studies, an indigenous practice to separate different social spaces.[17] Furthermore, the round shapes indicate African houses, both for freemen and captives, while the squares stand for straw huts predominantly occupied by the Company's European employees.[18] The observation by La Courbe is confirmed by the mixing of squares and circles on the maps within the confines of the *tatas*. Only one straight row of European huts breaks the complex geometry of this conglomeration.

[88]

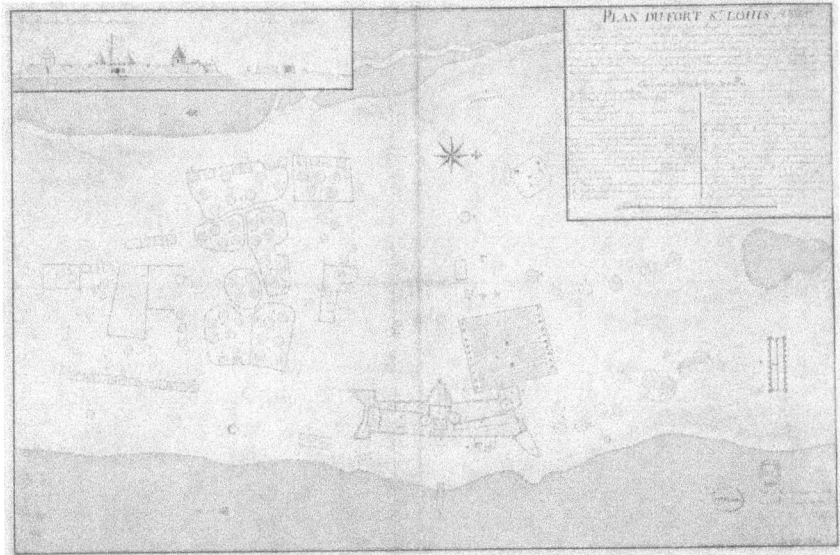

26 François Froger: Plan du fort Saint-Louis, 1705 (BNF Paris, Département des cartes et plans, GE DD-2987 (8127 B)).

The north-eastern bastion is described as having partly collapsed. Froger attributed that to the sabulous base the castle is built on, which made it necessary to gradually rebuild all the other walls, too. The repairs would have occupied all the Company's employees, including not only its workers, but also its slaves, and the laptots who received salaries. Other buildings included all the necessary installations for the colonial settlement: limekiln, forge, workshops for carpenters, a dovecote, and some huts for the storage of merchandise, placed some distance away from the settlement due to the risk of fire. A feature that was characteristic for the Senegal region, however, was the boathouse for the craft of the Senegalese laptots, the indispensable experts for all transportation to and from the island.[19] The 'governor's' mansion, however, was the largest building within the fortress, but did not feature any representational design. A large white flag on a wooden mast counts as the sole emblem of French colonial rule over the little outpost.

Another undated map, archived in the Bibliothèque Nationale as being from roughly the same period, shows the same spatial order on the island (see Figure 27). It shows a larger view of the islet than Froger's map, but also shows the settlement in more detail. The houses for the Bambara slaves were situated to the north of the fortress. In the settlement where the Europeans and free Africans lived together there

[89]

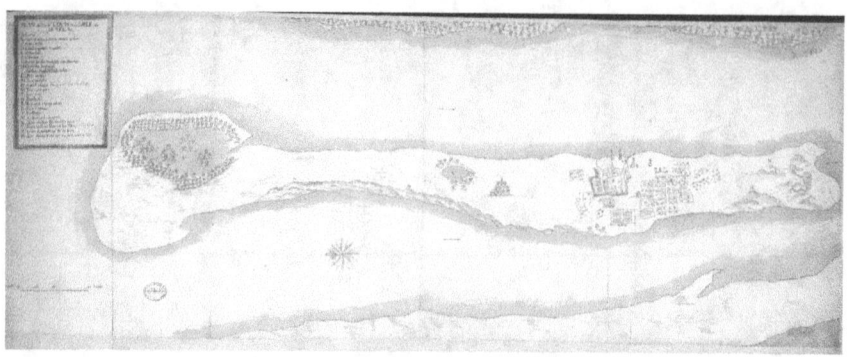

27   Anonymous: Plan du fort Saint-Louis et de l'isle du Sénégal, s.d.
(BNF, Département des cartes et plans, CPL GE DD-2987 (8126 B)).

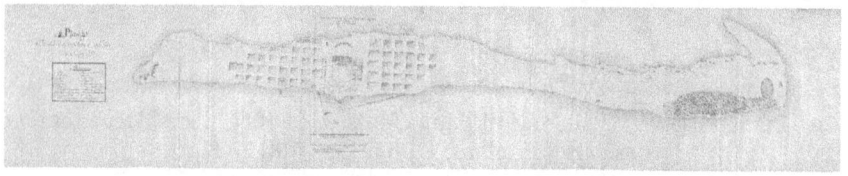

28   François Froger: Dessein du Fort propose à faire sur l'Isle du Senegal,
25 November 1704 (CAOM 19DFC/83).

were three contained groups of houses assigned to three men: a François
Le Clerc, Etienne Delarue, and a M. de Bosz, who are designated as
habitants, the usual name for European settlers. This confirms the
impression of the presence of a mixed-living culture on Saint-Louis
Island.

Froger's plan for the proposed new fortress, therefore, was intended
to impose a more dominant order on the French-African settlement
(see Figure 28).[20] He designed a fortress that was nearly eight times
larger than the old building. Its polygonal shape with four bastions
covered the whole width of the island. To the north, a grid system of
streets was reserved for the African houses; the Eurafrican mixture was
thus removed. The European workers, soldiers, and sailors were to be
accommodated in a large two-storey building by the northern wall.
The opposite building to the south was reserved for officers, more
important merchants, Company agents, but Senegalese laptots (maîtres
de barque), too.

The realization of the project seemed difficult even to the engineer
himself: 'The project of the fortress would seem extraordinary even to
those who don't realize they will have to transport all the building

materials (*moilon*) via the sea from Gorée Island, which is nearly forty leagues away from Senegal, and that there is nothing but sand and virtually no soil on the Senegal Island.'[21] Froger intended to solve this difficulty by 'treating the terrain with care' (*ménager le terrain*) and to limit the dimensions of the castle to a moderate size. Perhaps this is a concession to the unemotional design of the new fortress and the lack of material and financial resources that did not allow for a grander castle. This plan, however, was also not realized.

In 1723, a *mémoire* reported that the fortification of Saint-Louis did not deserve much of a mention. It had, however, thirty cannons placed in its batteries, with sixty Europeans and forty Bambaras (African slaves captured in the upriver hinterland) operating them. Furthermore, the bar at the outlet of the Senegal River made all other fortification obsolete.[22] Gorée had better fortifications, with forty cannons in place and forty European and forty Bambara soldiers. Of greater concern was the only stronghold in the Upper Senegal Valley, in the kingdom of Galam, where the Company maintained a small fortress at the Falémé River. Fort Saint-Joseph was very small and armed with only twelve cannons and fifteen European and fifteen Bambara soldiers. It was built simply of clay, and the holding cells for slaves were very small.[23]

Fort Saint-Joseph (built 1700, destroyed 1701, rebuilt 1714) is an interesting case of French building effort, since it took a long journey by boat via the Senegal River to travel the distance to the next entrepôt in Saint-Louis.[24] The dependence on African contractors, navigators, workers, and artisans was even greater than elsewhere. In order to strengthen the defences and to build larger holding cells, materials and workers had to be transported to Galam aboard 'double sloops' navigated by laptots only in those seasons when the river carried sufficient water to get over the rapids. All the effort was worth it, the Company agent argued, since it opened up trade with the slave merchants of the far inner reaches of the country.[25]

In 1714, Fort Saint-Pierre was built further upstream on the right bank of the Falémé River. Its purpose was not so much to trade in slaves, but rather to secure access to the gold mines of the kingdom of Bambouc. There was also a plan to erect a third fortress in this region, but it was not developed beyond a simple wooden structure. According to Labat, slaves from Bambara were instructed to perform the carpentry and to excavate a ditch in the form of a square or pentagon. An all-powerful commander directed the construction, Labat writes, which led to a short flourishing of French trade in this region to the detriment of Mandinga, Portuguese, and 'Angolan', that is, Lusoafrican merchants.[26] According to the list of furniture, Labat adds to his account, the French officer maintained an almost opulent lifestyle. Besides basic furnishing like matresses, cushions, and bed linens, the three fortresses were equipped with tablecloths, napkins, aprons, towels, tin glasses, and cutlery, as well as hourglasses and compasses; the commander even provided for the parts of a carriage, including the oil paint for coating the coach.[27]

Fort Saint-Joseph, after being destroyed several times by floods, was rebuilt once more after 1782 in a more rectangular shape and at a more secure location than the original one. The pragmatic artillery officer Pierre-François-Gabriel Destauches, also an engineer-geographer previously posted in the Antilles,[28] proposed to buy land from Heylimar Boudou, the ruler of the land between the two banks of the Falémé River, neighbouring the kingdom of Galam.[29] To cut costs, the officer laid a plan before the minister to employ predominantly native workers. Even skilled workers could be procured from among the Africans in Senegal: one blacksmith, one cabinetmaker, two carpenters, two master masons, four assistant masons, twenty workers or sailors from Saint-Louis, and twenty local workers from Galam. In total, there would be fifty workers, each paid five bars of iron each month over a period of eighteen months. The costs would amount to 9,000 livres tournois for the salaries, 2,700 livres for provisions, and total 17,550 livres after setting aside another 5,850 livres for unforeseen expenses. Although the surrounding area could provide most of the building materials, Destauches requested 2,000 planks of fir wood, this being more durable than that of the native trees. He also asked for 2,000 bars of iron, window casements, door hinges, locks, 400 pounds of nails, and various tools for cabinetmaking and forging.

The defensive purpose of these fortresses in Galam was directed against the native people and against the invasions of the nomadic Moors from north of the river, but not, as in most other cases in the early modern colonial fortification practice, against other European nations. Together with the trading posts in Farbana and Samarina, the fortresses made up a loose network, forming a protocolonial contact zone that already demarcated the limits of the territory in French West Africa as it was occupied in the nineteenth century. Two fortresses in Podor and on Caignou Islands along the Senegal River served as relay stations on the route to the Atlantic coast. But far from dominating, the French strongholds were able to be built, maintained, and defended merely with African collaborators. Fortresses and indigenous settlements were interconnected through trade and the exchange of merchandise, materials, and both free and enslaved people.[30]

Whereas the fortresses of Galam did not last long enough to establish durable links between the French and the local people, the lively building activity on Gorée Island enables better observation of the formation of an emotional community. The style that emerged here, originally labelled 'African Rococo' by Mark Hinchman, had an influence over Saint-Louis, too. The Creole-colonial environment of Gorée was signified by Europeans and Africans, including people of mixed race, free blacks, and slaves, who 'lived in the same houses, in the same rooms, and sometimes slept in the same beds'.[31] Such a flexible architecture on the island, Hinchman argues, created 'a society where gender, race, and nationality were not categories that always denied people opportunities' and thus contributed to the formation of a specific African modernity.[32]

This mixed architectural culture that emerged on this small island absorbed Portuguese and French traditions as well as those of the local Wolof population (Waalo, Jolof, Kajoor, Bawol) and, perhaps most visible, the style typical of the architecture of the Mande people. The latter, also known as Bambara, were predominantly the slaves that were employed at the construction sites on Gorée and influenced the appearance of the buildings. Perhaps most prominent in this regard is the Maison Pépin (otherwise known as the Maison des Esclaves) that featured a style that was the result of reciprocal influence of different European and African architectures (see Figure 29). Pilasters

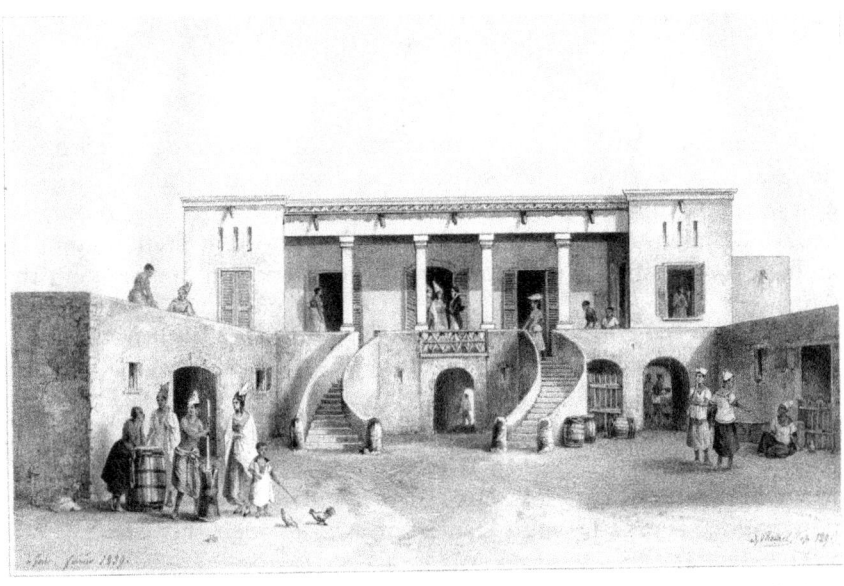

29 Adolphe d'Hastrel: Une Habitation à Gorée (Maison d'Anna Colas), 1839: Maison Pépin. Today the iconic Creole house is called Maison des Esclaves – House of Slaves; it is subject to debate, however, if this building served in fact for the imprisonment of slaves.

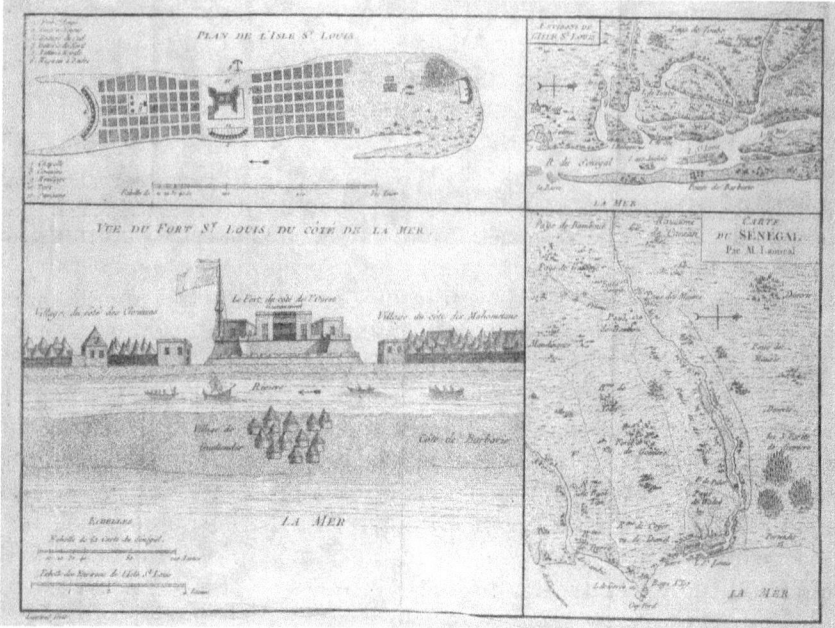

30 Plan de l'isle St. Louis. Vue du Fort St. Louis du côté de la mer.
Environs de l'isle St. Louis. Carte du Sénégal, in Lamiral, *Illustrations de
l'Afrique* (BNF, Réserve DT 549.8 L 23).

with capitals, dentils, regular fenestration, and overhanging cornices,
for example, were attributes typical of houses in Timbuktu in the
east of the Mande country. Its iconic staircase with its clay rendering
evokes Saharan aesthetic themes as well as those of French colonial
architecture, pointing, for example, to the plantation mansions in the
Antilles.[33]

A similar mixture of styles can be observed in the improvements to
the government building in Saint-Louis that were conducted during
the eighteenth century (see Figure 30).[34] Around 1800, the buildings of
the town were eventually ordered according to the colonial grid pattern.
A plan of the island from 1789 shows a confessionally divided African
town with one Muslim and one Christian quarter. In the middle, the
newly styled Fort Saint-Louis stands out, with by then only two bastions
to the west and two wings to the east.

A lateral view from the same sheet shows the west façade of the
fortress, revealing the architecture of the government palace inside. It
is a two-winged building with a flat roof, large vertical windows to

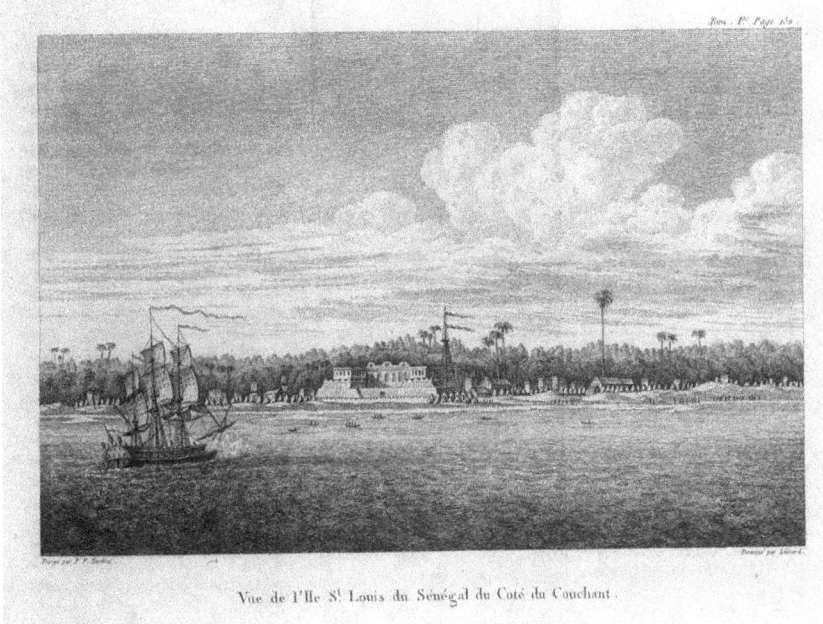

Vue de l'île St. Louis du Sénégal du Coté du Couchant.

31 Vue de l'île St. Louis du Sénégal prise du côté de la mer, in Geoffrey de Villeneuve, *Illustrations de l'Afrique*, vol. 1, p. 63 (BNF Paris, Réserve DT 549.2 G 34 v1 et v2 A).

each side, and a broad balcony supported by two columns. The style is distinctly not inspired by contemporary French architecture, but is similar to the Lusoafrican architecture that is so prominently on display on Gorée Island.

In 1814, a print in René Claude Geoffrey de Villeneuve's *Illustration de l'Afrique* shows Fort Saint-Louis also as viewed from the sea to the west (see Figure 31). It is a slightly different depiction, with the flagpole at the other side and the two wings defined by a row of several windows, but no balcony. Instead, there is a balustrade on the flat roof with three trapezoid pinnacles. It is possible that the designers of these sketches have mixed up the perspective, because the caption 'vue prise du côté de la mer' on both prints is ambiguous, since the Senegal River is quite broad at this location. Be that as it may, the style of Fort Saint-Louis on both images is not French, but distinctly local. The geometric ornaments appear regularly on traditional West African architecture, for example, in the buildings of the Toucouleur people, a Muslim ethnic group that settled in today's Mali, Mauretania,

[95]

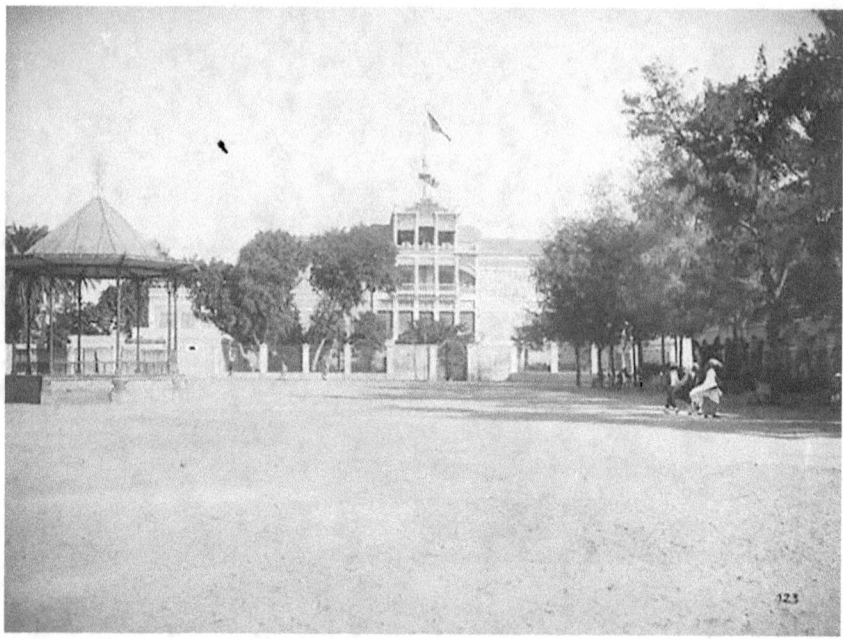

32 Anonymous, Hôtel du Gouvernement, Saint-Louis, Sénégal, 1900
(CAOM 30Fi26/1).

and Senegal. Their houses feature porches with wooden columns and
adobe-style façades with glassless openings in the walls in triangular,
square, and trapezoid shapes.[35]

During the nineteenth century, the government building was
stripped of its fortification and became the Hôtel du Gouvernement.
The layout of the building remained the same; a tower with a flagpole
and a veranda with balustrade on the roof were added before 1900.
Contemporary photographs show that the original features persisted
throughout the transformation of colonial French West Africa. In the
centre, the façade with large vertical windows regularly divided by
columns made reference to the eighteenth-century African design of
Lamiral's collection of illustrations (see Figure 32). Another image
shows the two wings of the building with additional balconies on
both sides imitating the traditional porches of Toucouleur and
Portuguese-styled houses. The Palais du Gouvernement still stands
in the centre of the old town of Saint-Louis. Today, its façade has lost
most of its Creole appearance – European ornaments dominate, but the
lofty structure survives and is testimony to the mixed history of this
structure.

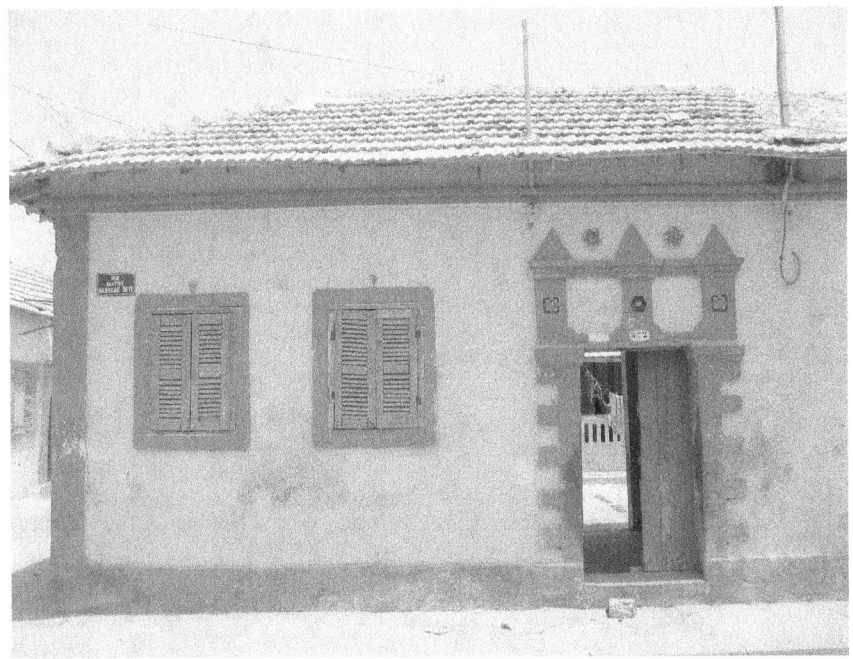

33 Bart van Poll: Creole House in the Rue Maître Babacar Sèye, Saint-
Louis, Senegal, 2006.

Notes

1 For the general history of the region, see Prosper Cultru, *Histoire du Sénégal du XVe siècle à 1870* (Paris: Emile Larose, 1910), which is still the most complete account. See also André Delcourt, *La France et les établissements français au Sénégal, 1713–1769*, Mémoires de l'Institut français d'Afrique noire, 17 (Dakar: IFAN, 1952); Abdoulaye Ly, *La Compagnie du Sénégal* (Paris: Présence africaine, 1958).
2 Benjamin Steiner, 'The Monuments of Empire: Global Material Culture, "Colonial" Spaces and Emotional Styles in French Senegambia (c. 1630 – c. 1730)', *Cromohs* 20 (2015), 52–76; see also Toby Green, 'The Emergence of a Mixed Society in Cape Verde in the Seventeenth Century', in Toby Green (ed.), *Brokers of Change: Atlantic Commerce and Cultures in Precolonial Western Africa*, Proceedings of the British Academy, 178 (Oxford: Oxford University Press, 2012), pp. 217–36.
3 Peter Mark, *"Portuguese" Style and Luso-African Identity: Precolonial Senegambia, Sixteenth–Nineteenth Centuries* (Bloomington, IN: Indiana University Press, 2002); George E. Brooks, *Eurafricans in Western Africa: Commerce, Social Status, Gender, and Religious Observance from the Sixteenth to the Eighteenth Century* (Athens, OH: Ohio University Press, 1998).
4 Mark Hinchman, 'African Rococo: House and Portrait in Eighteenth Century Senegal' (PhD Dissertation, University of Chicago, 2000); published as *Portrait of an Island: The Architecture and Material Culture of Gorée, Sénégal, 1758–1837* (Omaha, NE: University of Nebraska Press, 2015).
5 CAOM Col C6 1, Etat de l'habitation du Sénégal, 1664.

6 CAOM 16DFC/82, No. 4: Ducasse: Mémoire ou Relation du S. du Casse sur son voyage de Guinée [...], 1687–1688, p. 3: 'Les Portuguais ont esté les premiers qui l'ont connue cy l'année 1443, et sy s'establissent en 1448; les François y aller eut negotier envirn en 1455, et les negres assurent qu'il y a plus de 200 ans qu'ils en sont pocesseurs.' The Mémoire is edited and published by Paul Roussier (ed.), *L'établissement d'Issiny, 1687–1702* (Paris: Larose, 1935). For the early Senegal Company, see Benjamin Steiner, 'La première Compagnie du Sénégal de Rouen de 1633', in Éric Roulet (ed.), *Le monde des compagnies: Les premières companies dans l'Atlantique*, vol. 1: Structures et modes de fonctionnement (Aachen: Shaker, 2017), pp. 145–59.

7 CAOM 16DFC/82, No. 4: Ducasse, Mémoire, p. 4: 'Jusques a Serrlionne, commerce annuel qui sy fait, 10000 cuirs, 130000 livres de gomme arabique, 20000 livres d'ivoire, 200 esclaves, 2 à 3 quaisses de plume d'Autruche, 3 à 4 livres d'or, l'ambre gris y est esté comun autrefois, mais il y a 20 années qu'on y a veu.'

8 Ibid., p. 108.

9 Ibid., p. 110: 'Cette depence est de tres peu de chose y ayant sur le lieu des pierres et de la chaux.'

10 CAOM 16DFC/82, No. 6: La Courbe: Mémoire relatif à l'attaque des forts d'Arguin, Sénégal, Gorée, Gambie, Serraléone et Madrabombe, 1693, fol. 1r–v.

11 Ibid., fol. 4r.

12 CAOM 16DFC/82, No. 5: Mémoire de l'estat ou estoit le Senegal avant qu'il fut pris par ses ennemis de l'etat et ce qu'il faut faire pour le reprendre, 1693, fol. 2v.

13 CAOM Col C6 1, Chambonneau to Seignelay, July 1688.

14 Michel Jajolet de La Courbe, *Premier voyage du Sr. de La Courbe fait à coste d'Affrique en 1685*, ed. Prosper Cultru (Paris: Champion/Larose, 1913), pp. 24–6.

15 CAOM 18DFC/7C: La Courbe: Croquis de l'île de St. Louis et des environs, 1694.

16 Francois Froger, *Relation d'un voyage fait en 1695, 1696 et 1697 aux côtes d'Afrique, détroit de Magellan, Brésil, Cayenne et isles Antilles, par une escadre des vaisseaux du roy, commandée par M. De Gennes* (Paris: de Fer, 1698); Francois Froger, *Relation du premier voyage des François à la Chine fait en 1698, 1699 et 1700 sur le vaisseau L'Amphitrite*, ed. E. A. Voretzsch (Leipzig: Asia Major, 1926).

17 Ibrahima Thiaw, 'Atlantic Impacts on Inland Senegambia: French Penetration and African Initiatives in Eighteenth- and Nineteenth-Century Gajaaga and Bundu (Upper Senegal River)', in J. Cameron Monroe and Akinwumi Ogundiran (eds), *Power and Landscape in Atlantic West Africa: Archaeological Perspectives* (Cambridge: Cambridge University Press, 2012), p. 61.

18 See the legend on Froger's map: '5. 5. Cazes de paille servant de logement à la plûpart des commis, ouvriers, et autres emploïez de la Compagnie | 6. 6. Cazes des Negres tant libres que captifs.'

19 On this independent group of indigenous logistics contractors, see Mark, *"Portuguese" style*, pp. 57, 88, 104–5; Brooks, *Eurafricans*, pp. 52–4.

20 See also CAOM 16DFC/82, No. 14: Explication des lettres et chifres, qui se trouvent au Plan et Profil du Fort propose faire sur l'isle du Senegal, 25 November 1704.

21 CAOM 16DFC/82, No. 14: Explication, fol. 1: 'Le projet de ce Fort paroistra extraordinaire à ceux qui ne sauront pas qu'il faut faire venir de l'Isle de Gorée, qui est à prés de 40 lieues du Senegal, tout le moilon par mer, et qu'on n'a que du sable et Presque point de terre dans l'Isle du Senegal.'

22 CAOM 16DFC/82, No. 18: Mémoire sur le Commerce du Senegal, 14 October 1723, fol. 2v.

23 Ibid., fol. 3r.

24 See Père Jean-Baptiste Labat, *Nouvelle Relation de l'Afrique Occidentale contenant une description exacte du Senegal & des Pais situés entre le Cap Blanc & la Rivière de Serreleone, jusqu'à plus de 300. lieues en avant dans les Terres. L'Histoire naturelle de ces Pais, les differentes Nations qui y sont répandues, leurs Religions & leurs moeurs. Avec l'etat ancien et present des Compagnies qui y font le Commerce. Ouvrage enrichi de Quantité de cartes, de Plans, & de Figures en taille-douce*, 5 vols (Paris: Guillaume Cavelier, 1728), vol. 4, pp. 25–102. Labat

reaches the conclusion that it is necessary 'to know the country in depth as well as the habits and to make friendships' with the indigneous people (p. 30). His account is, however, not as reliable as it is for the Caribbean since he had never been there and relied on a partly forged collection of relations by the director André Brüe to whom he attributed many accomplishments that were, in fact, achieved during the tenure of the director Michel Jajolet de la Courbe (cf. Cultru, *Histoire du Sénégal*, p. 73).

25 CAOM 16DFC/82, No. 18: Mémoire, 1723, fol. 3v–4r: 'On osera dire que ce comptoir bien fourny donnera a la Compagnie plus de mille negres, la commodité du Niger que les caravannes suivent assés volontiers por ne pas manqué d'eau dans leur chemin, y attire les marchands d'esclaves de l'intérieur des terres les plus eloigniées.'
26 Labat, *Nouvelle Relation*, vol. 4, pp. 63–4.
27 Ibid., p. 69ff.
28 See his dossier in CAOM COL E 130.
29 CAOM 16DFC/82, No. 62: Destauche: Mémoire sur l'utilité d'un établissement à Galam, 12 January 1782, here fol. 2v.
30 CAOM 18DFC/60A: Sarrazin de Monferrier: Plan du Cours du Niger, Grand fleuve du Sénégal navigable, ca. 1782.
31 Hinchman, *Portrait of an Island*, Introduction.
32 Ibid.
33 Ibid.
34 For this later period, see Hilary Jones, *The Métis of Senegal: Urban Life and Politics in French West Africa* (Bloomington, IN: Indiana University Press, 2013).
35 Hinchman, *Portrait of an Island*, Introduction.

5

Variegated engineering: The builders of the Caribbean empire

In theory, French colonial administrators were convinced, the art of engineering and building underlined the supremacy of their rule in the empire. They were the ones that improved cities, made them look more beautiful and, in short, modernized the colonial empire. This narrative, however, must be questioned when one takes a closer look to those experts and techniques that contributed to the effort to achieve a somewhat ideal order of a colony. In fact, the modernization of colonies was variegated as well as were the people that were involved in the projects to build the empire.

Franco-Afro-Carib origins

The group involved in the construction of fortresses, bastions, batteries, ports, piers, sea walls, and public buildings such as manor houses, churches, convents, and hospitals cannot and should not be reduced to French protagonists. From the beginning of colonization, there are several instances of collaboration, sharing of technologies and materials, and patterns of mutual reaction to the respective building styles. Starting with the building practices themselves, this chapter gives examples of this 'mixed economy' of engineering, which is perhaps the most characteristic trait of the French colonies in the Antilles.

First of all, it is important to list the *dramatis personae* that played their particular parts in this history. After the arrival of Europeans, the islands were slowly transformed into agricultural colonies, and infectious diseases and warfare led to the gradual displacement and eventual extermination of the indigenous population.[1] Until the middle of the seventeenth century the few remaining Amerindians were still a political factor on Martinique, Guadeloupe, Saint-Christophe, and the smaller islands of the Lesser Antilles, but their population was reduced to only a few thousand after the French led colonial wars against them under the governments of d'Esnambuc, Parquet, and Houël. In 1671 a census for Guadeloupe mentions only fifty-one 'sauvages, metis et tabouis'.[2] On Saint-Domingue none of the pre-contact Arawak population survived, and some indentured or enslaved Indians who were captured on raids still remained on Tortuga.[3] Some of the Carib people persisted as 'Black Caribs' (Garifuna), descended from marooned African slaves on St. Vincent (exiled again from there by the British to Roatan Island off Honduras in 1795), on St. Lucia, Guiana, and on Dominica, where the Kalinago, the only surviving indigenous population, continue to exist in small numbers up to the present day.[4]

But despite the decimation of Amerindians, the group was still exercising an influence on the colonial society and its infrastructure. The survivors of the native peoples, James Pritchard notes, 'continued to play important roles in the history of the French colonies: as warriors fiercely resisting colonial incursions, as partners in trade, as agents of imperial conflict, and as independent actors pursuing their own tribal policies'.[5] Even if this assessment applies better to the indigenous people in Canada, the case of the West Indies is not completely dissimilar. Especially during the seventeenth century, Caribs, as they had already been called on Blondel's map of Martinique, lived not only close to the French settlers, but were also part of the colonial building efforts.

African immigrants coerced to leave their homes and submitted to the transatlantic slave trade belonged to the second distinct group that shaped the social and economic structure of the colonies.[6] Perhaps Africans were the most influential actors in regard to the material transformation of the landscapes, large buildings, and the general appearance of settlements. The number of slaves in the French West Indies increased drastically around 1700: in Saint-Domingue there was a rise from 4,000 before 1680 to 176,560 imported slaves in the period to 1730, Martinique had 15,830 slaves up to 1679 and had added 91,260 by 1730, and Guadeloupe saw a similar rise from 10,000 in 1699 to 83,080 new arrivals up to 1730.[7] Despite wars, economic turmoil in France, and other difficulties in the colonial administration, the number of imported African slaves was more than ten times the number of all European migrants to French America, including Canada.[8]

Africans were not only the most important group in demographic terms, but also in the way they enabled the realization of projects set out by French administrators, proprietors, religious orders, and engineers. The overarching system was, of course, the slave economy, established on the islands by the first Europeans and vastly expanded in the course of the eighteenth century.[9] It coerced Africans to perform hard labour, necessary to keep up the sugar production, but also on the construction sites for the fortification of the islands. Also, Africans brought with them building techniques, architecture, and aesthetic conventions that were applied to the buildings they used for themselves and characterized a particular style of living.

Immigration from France to the American colonies was very limited, not only in comparison with that from Africa, but also compared to that of other European nations. It seems very few settlers left France to try their luck in the colonies: it is estimated that not more than 100,000 emigrants came to the West Indies and Canada, amounting to only 5 per cent of the overall European transatlantic migration between 1500 and 1800.[10] Despite the low demographic impact French people had on the colonial societies, their role has tended to be overestimated so far. Why this is the case seems unclear. Most of the settlers arriving in the colonies from France were neither qualified to contribute to the growth of the economy nor, for that matter, skilled enough to work as artisans on sophisticated projects like fortification. There were, as a first subgroup, a few immigrants who arrived as freemen – adventurers, planters, merchants, tradesmen, government officials, and soldiers – and formed, if they stayed, a white Creole society. Secondly, there were poor emigrants, pushed out of France due to economic hardship or, as in the case of the Huguenots, their precarious religious status. Usually, they indentured themselves to ship captains, planters, or merchants for three years' service and worked as *engagés* on the islands.[11]

The latter group was particularly relevant to the first effort to colonize the Antilles as they were put to the necessary hard labour. Most of them, however, were vulnerable to infections and parasites, which often made their experience traumatic. The migrants were often pressed onto ships, mostly in French port cities such as La Rochelle. Only sometimes were the *engagés* skilled workers. If they boarded a ship in France the royal administration ordered the captain to keep record of the disembarkation in order to get them to a public works project. These skilled workers rarely received formal contracts and, as impoverished newcomers, had difficulty integrating into the island's hierarchical society of slave plantations and small colonial towns.[12] Where large building projects were concerned, however, skilled indentured servants proved themselves a second backbone alongside the African workforce.

Indigenous models

When the first French settlers came to Martinique they were no more than a small group struggling to survive. But the economic spaces of the plantations gradually contributed to the emergence of a 'société d'habitation' characterized by a frontier mentality that entertained close relations with the Carib people.[13] The so-called buildings of these settlements did not constitute proper villages or towns, but rather were loose ranks of storehouses and huts, usually built of wood.[14] But these buildings underwent continuous change. According to Gabriel Debien, there was so much building activity going on that the islands resembled a permanent construction site (*chantier permanent*).[15]

The local building techniques very often exhibited traces of indigenous influence. The structure would consist of wooden posts driven three to four feet into the ground, enclosing a hearth on an earthen floor. The tops of the posts were bent in such a way that they formed a roof ridge and eaves. A wall of woven slashed reed or palm leaves surrounded the cabins, which were also covered with reed and sugar cane tops for protection against rain. This is, according to Père Labat, how most houses were constructed, at least until 1700, during his tenure on the island.[16]

The building site was prepared by cutting down trees without any regard to their size and then setting fire to the dry stumps. Thus, those trees suitable for construction were burned, too. The European inhabitants of Martinique, Labat tells us, copied the Caribs in this technique – a procedure the chronicler found to be disadvantageous to the island's environment and economy. Only a few people would conserve the large trees for making boards, beams, and other items and selling them for considerable profit. They would wait for certain moon phases to chop them down, keeping the timber in stacks under small shelters until they could be used. The burning of the smaller trees was also done in consideration of the strong winds and to cut aisles or paths through the remaining forests to control the spread of fire.[17]

Labat does not mention it expressly, but it is highly probable that these techniques of choosing timber and burning the rest of the forests may already have been used by the indigenous inhabitants and then adopted by the French. The negative connotation of the term 'bad habit' (*mauvaise coûtume*) suggests a Carib influence over the methods used to manage the island's environment and views on the advantages and disadvantages of transforming it for one's own purposes. If one reads Labat's text further against the grain like this it is possible to discover traces of the positive effects resulting from adapting to indigenous ways of life.

Another example of the undermining of indigenous agency is a passage from du Tertre. He casually admits their positive influence, yet denotes the Caribs as a savage menace for the settlements. While the houses of the wealthier French inhabitants had timber frames encased with planks, floors made of timber pilings or bricks, and a tiled roof, the less wealthy lived in huts enclosed only by a fence made of reed, 'particularly in places where one did not expect incursions by the savages'. The poorest, however, had cabins covered only with canes, reed, and Latan palm leaves, but, du Tertre adds, 'those [houses] are incomparably more pleasant than our thatched houses in France'.[18] The indigenous architecture also affected the living habits of settlers: while married couples slept in beds, others slept 'à la façon des Sauvages' on cotton mattresses, without any cushions, bed sheets, or blankets.[19]

It may be the case that the houses of the ordinary inhabitants of the French Antilles were not very elaborate because they only intended to stay there for a short period of time. The Creole historian Médéric-Louis-Élie Moreau de Saint-Méry reported at the end of the eighteenth century that settlers incessantly spoke of their intent to return to France and therefore did not care too much about their permanent residence.[20] But usually this was merely wishful thinking on their part and most of the French lower-class citizens had to stay on the islands for a long time. But most houses were of a light and impermanent composition and served primarily functional purposes, such as providing spaces for social life, storage, and rest.[21] Labat described his own residence, which local carpenters had constructed in his parish of Macouba on Martinique. This house had only one main room and one sleeping chamber, also used for storage. An attic was reached by means of a staircase, and here Labat installed several hammocks where he and his guests could retire to rest. The house was entered through two doors from either the garden or the courtyard up a stone staircase, and the courtyard at the front of the house was on a slight slope so that water would run off.[22]

African-Creole expertise

The architectural style of the minor buildings in the Antilles was predominantly African. Du Tertre manages, by means of a veiled Latin reference to Seneca's letters to Lucilius, to convey a sense of African agency. It is a topos from ancient literature that serves to compare their dwellings with those of the people of the golden age: living in houses with 'forked poles at either ends, [...] close-packed branches and with leaves heaped up and laid sloping they contrived a drainage for even the heaviest rains; beneath such dwellings, they lived, but they lived in peace'.[23] The similarity of the houses of African slaves with those of the people of the mythical golden age carries a degree of condescension, but also represents the tradition of seeing African culture through the lens of the European narrative of an ambivalent human progress. The quote, in fact, is incomplete; it continues: 'A thatched roof once covered free men; under marble and gold dwells slavery.'[24]

It may have been du Tertre's way of expressing regret for the condition of African slaves, even though he flips the comparison upside down in alluding to Seneca's criticism of his Roman contemporaries as slaves of their own luxury: they lived in marble houses, while the people in simple huts were freemen. In the case of the Antillean society it was the other way around: here slaves lived in simple cabins, while free and wealthy men could afford the stone houses. In any case, we learn from this on the practical level that the Africans of the Antilles lived in houses framed by furcated posts carrying a roof covered only with reeds (see Figure 34 for a modern example). The outer wall is made of thick wooden stakes without gaps between and not of reed, as is the case with the houses of the French, who prefer to have more air circulation inside. The African houses, on the other hand, are tightly enclosed for a number of reasons, but particularly, as du Tertre claims to know, because they do not want to be cold at night.[25]

Albeit they were not free, African families took responsibility for building their own houses. These were grouped together at a distance

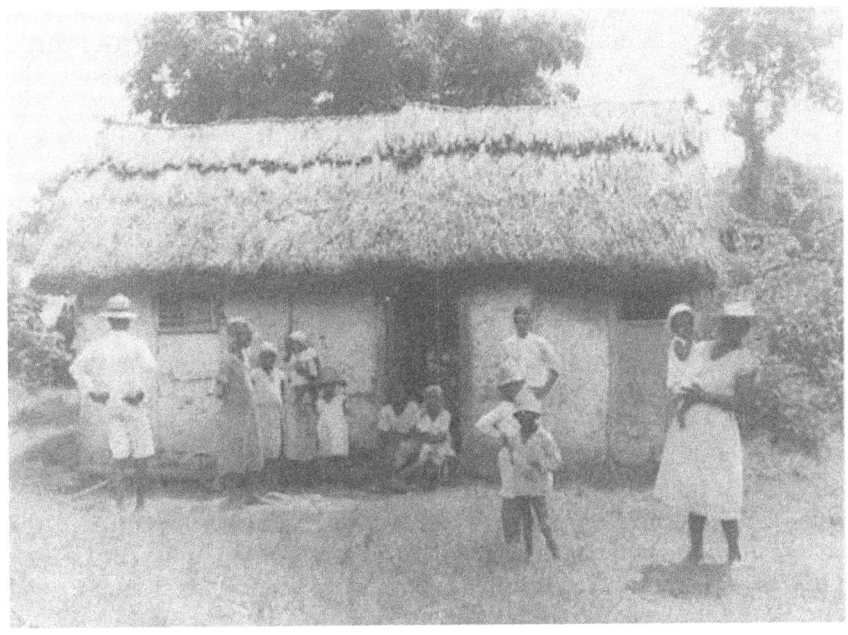

34 Robbez-Masson: Modern Type of a Workers' Cabin, Martinique, 1935 (CAOM 31Fi52/205).

of ten to twelve feet one from the other and sometimes forming a circle with a communal space in the middle. The governor Poincy enclosed the Ville d'Angole, where the Africans working for him lived in brick buildings, with a wall made of stone. Inside, the houses were of modest character, not furnished as richly as those of 'our savages', du Tertre remarks, but with cabinets and chests made by them.

One can infer from this that there were Carib and African construction experts mixing traditional building techniques either indigenous to the place or brought with them from Africa. Their contribution to other large building projects, such as that of the island's fortifications, was not merely that of a physical workforce, but consisted also in knowledge of technology, materials, and architectural and structural planning. This was, by the way, not the case when it came to the contributions of naval engineers to the building of houses for settlers on the islands. Only the church or rich individuals used the services of French fortification experts to lead some of their construction projects (see Figure 35).[26]

In the beginning of French colonization, artisans – masons, carpenters, or cabinetmakers – mostly originating from France had slaves 'of talent' whom they took into their apprenticeship and who then helped them in their work. But during the eighteenth century most of the island's artisans eventually were free black people.[27] Most of these African descendants still employed slaves 'of talent' at the construction sites. These black 'maîtres artisans' were, like their French colleagues, the most important experts for building projects, and it is probable that they even became acquainted with treatises on architectural theory that were usually the preserve of engineers educated in France.[28]

It is again Labat who gives us a lengthy but vivid account of Africans who work as experts.[29] Starting with a sugar refinery, where Labat employed several slaves himself, he describes in detail the different tasks that were performed by African men and women and the knowledge they required to do them. A sawyer, for example, was quite necessary because there was always a need for fresh boards, and Labat had several Africans learn this trade. But reading between the lines of Labat's somewhat narcissistic account it becomes obvious that Africans were, with or without the assistance of the Dominican, very able in performing all sorts of repairs. Labat's African carpenters were, for example, very capable of fixing the watermill that powered the saw, a complicated mechanism prone to failure. French carpenters, on the other hand, were often difficult to source for these tasks and sometimes reluctant to do the job.[30]

Having an African expert in the house, Labat tells his reader, is a treasure that one cannot value enough. On top of that, Africans were

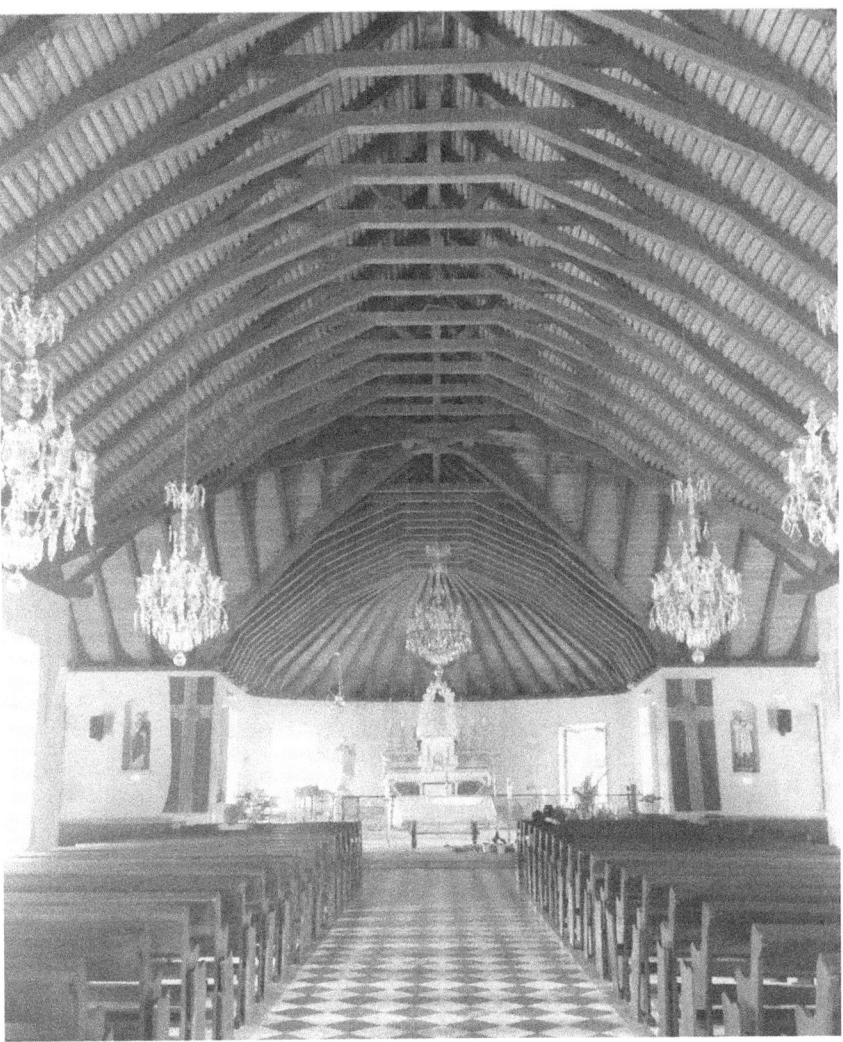

35 Wooden roof structure of the church Saint-Etienne in Le Marin, Martinique (built 1766).

very motivated to learn and to strive to become skilled workers in their own right. As experts, they felt themselves superior to other slaves. What Labat rather derogatively judged as mere vanity was actually genuine worker's pride: African masons or cabinetmakers even went to church with their aprons on and the rulebooks of their trade by their side.[31] From Labat's standpoint it made sense to utilize this hierarchy amongst the African population for his and other slaveholders' purposes.

To control the enslaved labour force he employed Africans, since 'there is no one in the world who commands a greater hold over them'.[32]

It comes, therefore, as no surprise to find that many African experts were employed in all sorts of tasks on the construction sites as well as in the wider logistics of the islands' fortification projects. Africans did not just perform the hard unskilled labour; they also had several occupations where they organized the work mostly for themselves. To produce mortar from lime, the workflow was divided between one group that gathered corals from the sea, which was the most important source of chalk in the Antilles (as it was on the coast of Africa), and another that was responsible for cutting wood to burn the lime. The quicklime was then loaded on board a boat for transport to the construction site. A commander, who could also have been of African descent, made sure that everybody stayed on task, probably by some means of coercion, and reported directly to the responsible engineer.[33]

Sources like the one quoted above allow only a glimpse into the everyday life of skilled and unskilled African workers on the construction sites of large building projects. It is, however, noteworthy that the enslaved people are mentioned in administrative correspondence even though they were not paid – only the amounts paid to their owners are specified. They were paid a daily rate for the work carried out by the slaves that were sent to the public works from the plantations. It is unclear, however, whether the the slave owners were also paid for the work of the teams at the limekiln, or whether these people received direct payments. That the latter was possible is shown by the letter of La Varenne from 1717 already mentioned above. African workers, like masons, could indeed earn money for themselves, even if they were paid less than European artisans.[34]

It remains uncertain as to whether Africans were regularly paid directly for their work or whether this was simply an exception. Most documents refer to expenses in respect of African slaves that were paid to their masters. On the other hand, one should not underestimate the role of those Africans who worked as lime fishers, lime burners, woodcutters, or boatmen transporting goods between the islands and archipelagos of the Antilles. In a letter from 1733 the engineer Houel complained about a stoppage of work at the construction site of Fort de la Basse-Terre (Fort Saint-Charles) on Guadeloupe caused by the lack of lime, sand, and rocks.[35] In order to prevent such logistical delays he underlined the importance of maintaining the sea transport system and those operating it:

I have also left the necessary instructions for amassing sufficient rocks and sand to finish the demi-bastion, with orders to burn the little limestone that remains while waiting for the boats to bring more. Since there are only three boats left of the thirty or so that are required, it is necessary to get hold of those [boats] that used to deliver [materials] regularly every year, since if these forced labourers [*corvées*] are neglected, as they have been since 1725, the construction works will definitely become considerably more protracted.[36]

Despite the ambiguity of the documents in respect to the merits of non-European workers it seems that the colonial administration and the leading engineers were quite aware of the importance of this infrastructure. There are strong indications that those 'entrepreneurs de la chaue', that is, those who managed the production and the transport of crucial building materials, were, in fact, Africans.[37]

This claim may be supported by Labat's account of his journey from Saint-Pierre to Fort Royal on Martinique. To reach this destination, and in order to avoid the mountainous route over land, he chartered a boat from a free African called Louis Galère, a name (*galère* means 'galley') obviously derived from the profession he had already pursued for two or three years. Labat informs the reader of his impression that this business was a thriving one, to the extent that others had imitated him and made 'not a small fortune'.[38] The charge was 1 écu per person or 6 écus for the whole boat; servants or slaves accompanying the traveller went free of charge. The ship-owner apparently did not operate the boat himself. An African employee would steer it, with four or five others rowing when there was not enough wind to sail. The boatmen piloted the vessel very ably through dangerous waters and were knowledgeable about weather conditions and where to make landfall in case of rain or storms during the journey.[39]

According to Hayot, the Sovereign Council of Martinique (*Conseil souverain*) taxed patrons like Galère heavily. They were often based in Saint-Pierre. Only after 1786 are there lists for Fort Royal, including five Africans, one câpre, and two mulattos.[40]

Africans, therefore, must be considered well beyond their contribution in terms of mere muscle power in the effort to complete large building works. They were crucial in performing skilled tasks and they were indispensable in providing the logistical support so necessary for an uninterrupted workflow in the building process. However, the range of materials they produced and delivered to the different construction sites is even larger, more diverse, and more complex.

Atlantic-Caribbean materials

The material resources of the Antilles did not provide everything needed for the realization of ambitious colonial building projects. It is quite remarkable that du Tertre devotes a whole chapter solely to stone in order to point out the importance of the aesthetic appearance of the local building material. There are stones, he writes, of a distinct silvery colour that changes over the course of the day into an opal colour, 'quite different from those I have seen in France'.[41] They could be found, for example, on the north coast of Basse-Terre on Guadeloupe (today's Quartier de Sainte-Rose), where the small fortress Saint-Pierre was erected by Charles Liénard de l'Olive after 1635. A certain green stone, perhaps a kind of silt, was considered by du Tertre not to be indigenous. Migrants from the mainland of South America near the Amazon River, he suspects, brought it to Guadeloupe. These hard stones were sculpted into figures, but only crudely, proving, as du Tertre adds, that the Caribs did not possess the proper tools to work this material.[42] That there were in the beginning no more houses on the island that were built of stone, but mostly huts of wood and reed, was not caused by the lack of materials, then, but rather a lack of experts. Du Tertre emphasizes that the islands were ideal for stone architecture due to the quantity of rock available that was suitable for stonecutting (see Figure 36 for a typical example of the island's masonry).[43]

The same was true also for making bricks and tiles since clay was ubiquitous. Plaster, however, was only found on the Îles des Saintes, located south of the Basse-Terre of Guadeloupe and brought there in 1641. Du Tertre found its quality similar to that in France, and even compared its application by the plasterer de Mouy on Guadeloupe with his masterwork at the Louvre! According to him, chalk was obtained from a white rock from the sea (*pierre marine blanche*), which was actually coral, and was just as good as the chalk used in Europe (see Figure 37). There were rivers that yielded pumice stones, too, which were, of course, produced by the volcanic activity on the islands, especially on Martinique. But there was, du Tertre claims, none of the variety of pebble or flint stones that could be found in France – a red stone found in some coves was the only one suitable to be used as an ignition source, especially needed for the flintlocks of muskets.[44]

Timber was the second major resource for building that was already present on the islands. Père Labat gives us a detailed account of the different sorts of wood suitable for carpentry.[45] The tropical hardwood was considered by Labat to be of a very high quality and very well suited for durable large structures, providing one had the proper equipment and qualified experts to work with it. It was difficult to saw this wood into planks without damaging the saw. Those who knew their profession sprinkled water on the saw to keep the rubbery fluid oozing out of the timber, but as this was hard work softer wood was sometimes used – a sign, in Labat's view, of the chronic laziness of the workers.

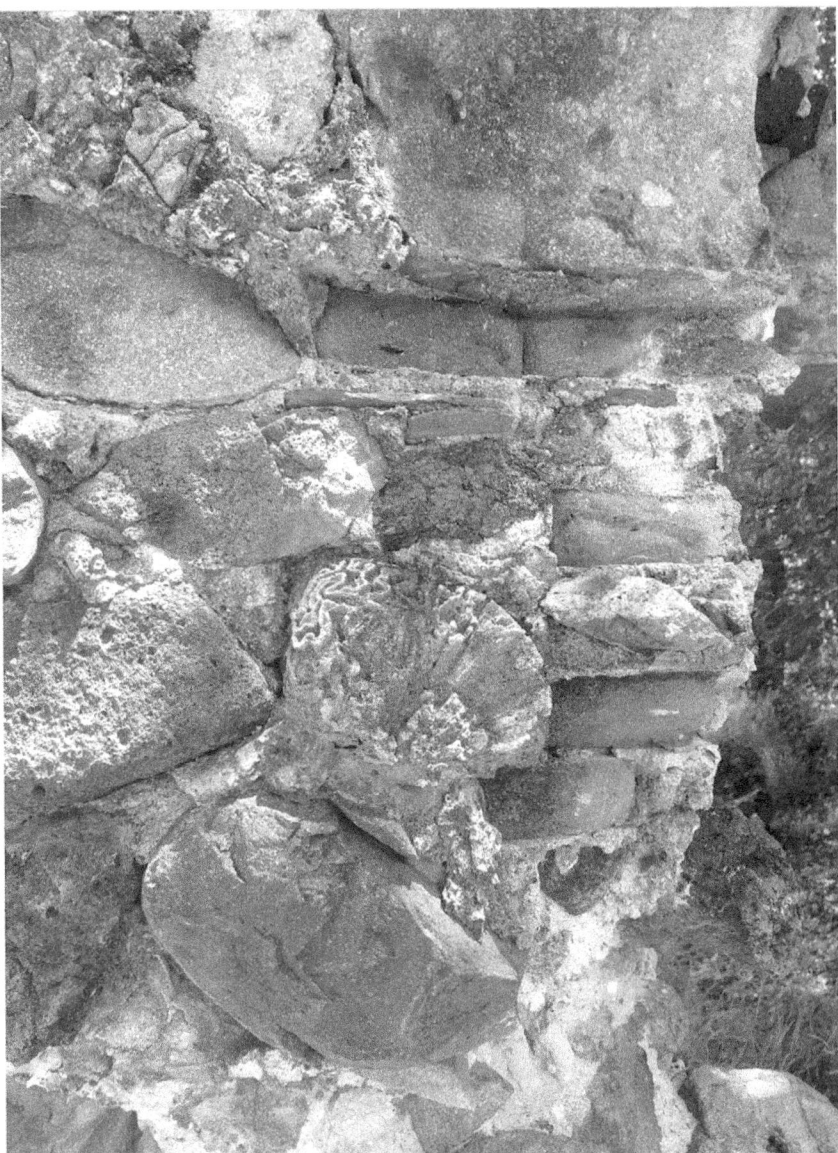

36 Typical mix of materials in a wall including bricks, boulders from the
vicinity, stony corals (*sceleractinia*), and lime mortar made from corals at a
building of the Chateau Dubuc (built 1721) on the Caravelle peninsula of
Martinique (for a short archaeological description, see Catherine Chasseur:
Chateau Dubuc, Trinité, Martinique, Master Thesis, Université Antilles-
Guyane, 1986, p. 38).

[111]

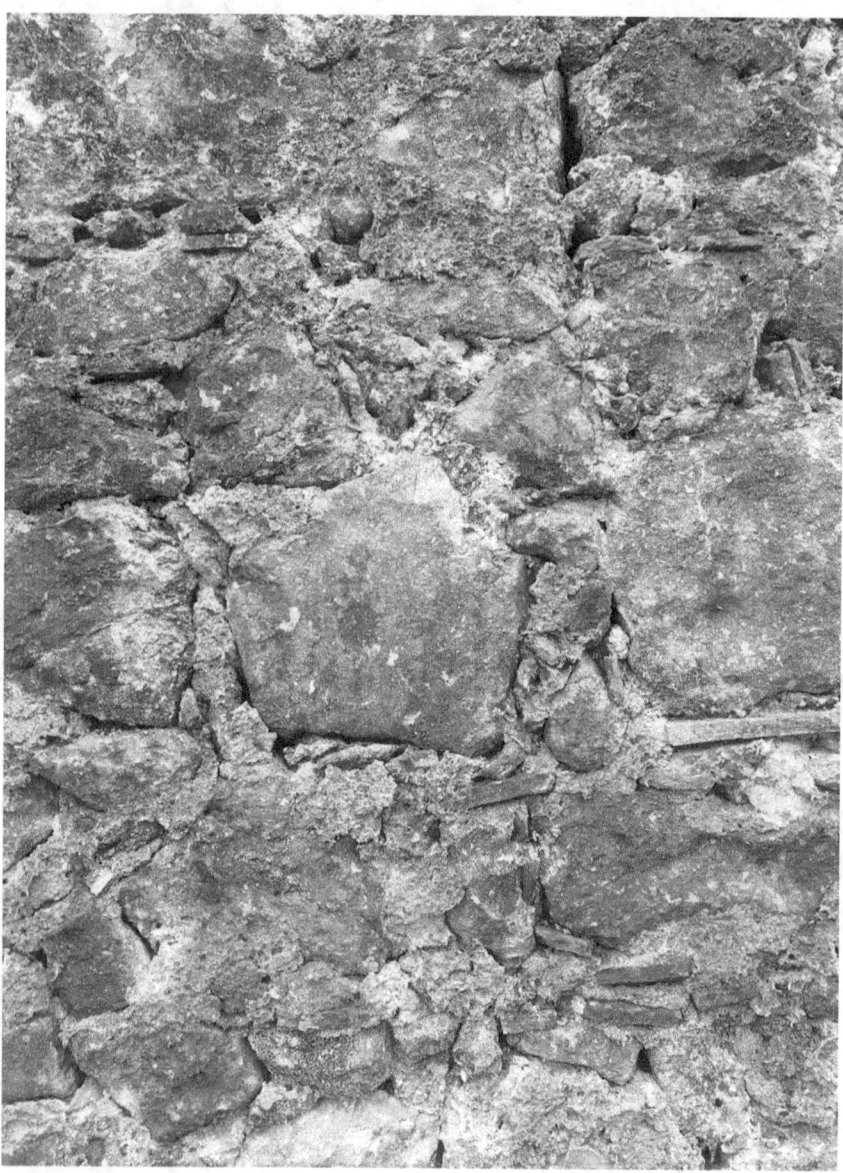

37 Detail of the wall of the Chateau Dubuc showing the black coating of volcanic stones and the coral lime mortar, usually mixed with cow dung and sugar syrup (*sirop de batterie*) to give it a whitish appearance (cf. Chasseur, *Chateau Dubuc*, p. 38).

The best trees, however, quickly disappeared because no care was taken to manage the forests sustainably. Labat tried to order the alternatives from amongst the indigenous population of trees according to their usefulness. Carib wood (*bois caraibe*), for example, was hard and dry but heated up in the sun, as a result of which he deemed it more suitable for interiors. Wood from thorn trees was used too, but mostly by cabinetmakers. Balatas trees provided the wood for larger beams, but the Acomas, the 'king of trees for building' (*le roi des arbres à bâtir*), provided the best material.[46] It was easily worked, suitable for polishing, and quite rigid but stable enough not to burst or crack. Lighter wood was used for battens or planks, mostly from the trees alongside the rivers (*bois de rivière*) or from the mountains (*bois de montagne*). Labat further mentions lianas, which were used to fasten the walls of the cabins, and the practice of covering the roofs with reed or sugar cane.

Even if many of the main building materials could be either obtained on the islands or produced on site, the accounts of Labat and du Tertre remain overly optimistic that these resources were sufficient for the larger building projects. In 1680 the engineer Payen, responsible for the work at Fort Royal, wrote that he not only needed more qualified masons from France, but also stone for the masonry itself, since that which was procured on the island itself was too expensive (see Figure 38). The price of one square toise of masonry was at 60 livres the previous year, which appeared to him excessive, and he wished to reduce the cost by examining options for obtaining materials from more distant places. Meanwhile, he used stone found in the escarpment of the site, which halved the cost. Nonetheless, he had to wait for the arrival of a ship from Rochefort in order to transport timber and limestone from other parts of the Antilles to Fort Royal.[47]

In 1765, the engineer Henri Philippe Joseph de Rochemore, who was responsible for the monumental development plan of Fort Saint-Charles in Basse-Terre on Guadeloupe, ordered diverse materials that had to be imported, among them fir wood (*bois de sap du nord*) and long-lasting wood (*bois incorruptible*), probably tropical hardwood.[48] Four years later, his successor Charles Le Beuf requested 50,000 *briques de provence*, 6,000 tiles, and 300 pounds of German steel (*acier d'Allemagne*).[49]

According to Charlery there were local brick potteries everywhere on the Antilles, but the quality of bricks formed in the shape of sugarloaves was rather mediocre and some materials deteriorated over time. The above-mentioned Poincy imported bricks as well as tiles from Holland to cover the floor in some of the rooms of his famous Château de la Montagne.[50] But this was probably an exception. Transporting building materials from Europe by ship was not necessary and

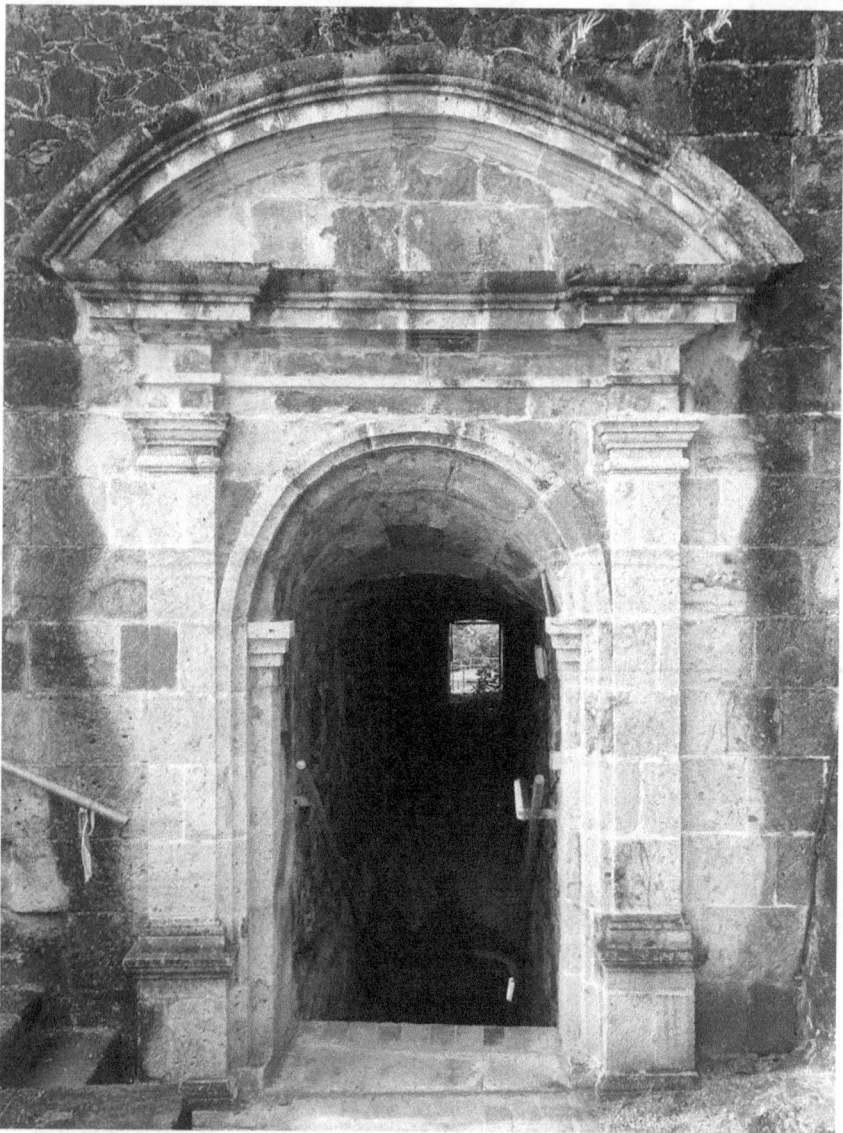

38 The 'Porte Louis XIV' inside Fort Royal, Martinique, showing the
recently sandblasted façade of lightly coloured stones that stand out in
contrast to the black rock walls of the fortress. The portal was constructed
in the beginning of the seventeenth century under the direction of the
engineer Caylus.

probably too expensive. Thus, the main resources were local and included many of the materials the indigenous population had been using decades before the arrival of the French.

On the other hand, the wish lists of items requested to be delivered from France to the Antilles reveals a much larger scope and variety of materials used in the colonial culture. Apart from stone, rock, and indigenous materials used for buildings, there were dozens of tools and special parts that needed to be imported from Europe. La Roulais, for example, ordered an extensive list of things, including 150 shovels, 200 fishing rods, 30 barrows, 36 hammers for the stonemasons, 12 angle metres, two grinding stones, 30 hatchets, 100 billhooks, 150 pincers, 36 mortar scoops, 20 needles, 3 old sails to build tents for the workers, and many other larger and smaller instruments.[51] The engineer also asked for basic building materials: 8,000 bricks, 1,000 brick tiles, and 7,000 tiles for the construction site at Guadeloupe. Whether this order was actually delivered to the Antilles is not certain, since the Navy Council delegated the order to the authorities in Rochefort.[52]

However, it can be inferred that materials from Europe were transported to the Antilles. Even basic items such as bricks and tiles had to be imported, either because the material made in Europe was of better quality, or because it was simply not possible to produce enough of them on site. The exact use of these materials is not certain. For example, were they applied to particularly prominent parts of the fortresses or other large buildings, for interior decoration or in different ways, to provide visual representation of European products? It is possible to speculate that they were used in a similar way to those ordered by Poincy, or alternatively, that they were simply employed in the same way as the local building materials. To gain a better idea of the function of materials, Chapter 7 seeks to reconstruct the contemporary impressions of buildings, of their materials, and, in a more aesthetic sense, their style.

Notes

1 Shelburne F. Cook and Woodrow Borah, *Essays in Population History: Mexico and the Caribbean* (Berkeley, CA: University of California Press, 1971), p. 401.
2 For some numbers, see Pritchard, *In Search of Empire*, pp. 8–10; Boucher, *Cannibal Encounters*, p. 35; Gérard Lafleur, *Les caraïbes des petites antilles* (Paris: Kathala, 1992), p. 21.
3 Pritchard, *In Search of Empire*, p. 8; Pierre de Vaissière, *Saint-Domingue: la société et la vie créole sous l'ancien régime, 1629–1789* (Paris: Perrin, 1909), pp. 74–5.
4 Louis Allaire, 'The Caribs of the Lesser Antilles', in Samuel M. Wilson (ed.), *The Indigenous People of the Caribbean* (Gainesville, FL: University of Florida Press, 1997), pp. 180–5.
5 Pritchard, *In Search of Empire*, p. 10.

6 The number of African slaves transported to the French colonies in the West Indies is particularly scarce for the seventeenth century; in the end of eighteenth century, however, France is considered to be the nation with the largest share in the transatlantic slave trade. See David Eltis et al. (eds), *The Trans-Atlantic Slave Trade: A Database*, http://slavevoyages.org, retrieved 17 May 2017, the most complete database that is continually updated. On African immigrants in the French Antilles, see Pritchard, *In Search of Empire*, pp. 11–16.

7 Philip D. Curtin, *The Atlantic Slave Trade: A Census* (Madison, WI: University of Wisconsin Press, 1969), p. 26.

8 Pritchard, *In Search of Empire*, p. 13.

9 For a general overview, see Laurent Dubois, 'Slavery in the French Caribbean, 1635–1804', in David Eltis and Stanley L. Engerman (eds), *The Cambridge World History of Slavery*, vol. 3: 1420–1804 (Cambridge: Cambridge University Press, 2011), pp. 431–49; cf. Gabriel Debien, *Les esclaves aux Antilles Françaises (XVIIe– XVIIIe siècles)* (Basse-Terre/Fort-de-France: Société d'Histoire de la Guadeloupe/ Société d'Histoire de la Martinique, 1974); Gabriel Debien, *Plantations et esclaves à Saint-Domingue* (Dakar: Université de Dakar, 1962).

10 Pritchard, *In Search of Empire*, p. 17; his estimate is that not even 10,000 French migrants settled in the West Indies during the period between 1670 and 1730.

11 Christian Huetz de Lemps, 'Indentured Servants Bound for the French Antilles in the Seventeenth and Eighteenth Century', in Ida Altman and James Horn (eds), *"To Make America": European Emigration in the Early Modern Period* (Berkeley, CA/ Los Angeles: University of California Press, 1991), pp. 172–203; Gabriel Debien, *Les Engagés pour les Antilles (1634–1715). La société coloniale aux 17e et 18e siècle* (Paris: Larose, 1952).

12 Huetz de Lemps, 'Indentured Servants', p. 199.

13 See Jacques Petitjean-Roget, *Les premières habitations de la Martinique. Monuments historiques, Architecture d'outremer* (Paris: Editions CNMHS, 1981).

14 Cf. Du Tertre, *Histoire générale des Antilles*, vol. 2, p. 449, who rejects the notion of calling the settlements towns since they would not fit the Aristotelian definition: 'C'est mal user des termes, ou ne pas sçavoir la définition d'une ville, qu'Aristote donne dans ses Politiques, comme font Messieurs Biet et de Rochefort; car il n'y a ny ville ny bourg, mais seulement quelques rangées de magazins bastis de pierres et de planches, où les marchands estrangers vendent ce qu'ils apportent, et où quelques artisans font leurs retraites pour la commodité au public, comme les tailleurs, le menuisiers, et autres semblables.'

15 Gabriel Debien, *Les Grand'cases des plantations à Saint-Domingue aux XVIIe et XVIIIe siècles*, Annales des Antilles, Bulletin de la société d'histoire de la Martinique, 15 (Fort-de-France: Société d'histoire de la Martinique, 1970), p. 14.

16 Labat, *Nouveau Voyage*, vol. 3, p. 48: 'Le terrein étant nétoyé on bâtit les cases ou maisons dont les poteaux se mettent trois à quatre pieds en terre avec une fausse sole. Le bout des grands et des petits poteaux est échancré pour recevoir le faitage et les sablieres. On palissade ou environne les cases avec des roseaux ou des palmistes refendus, et on les couvre avec des feuilles de palmistes ou de roseaux.'

17 Ibid., p. 46: 'La plûpart des habitans ont la mauvaise coûtume d'abbatre les arbres les uns sur les autres comme font les Caraïbes, et d'y mettre le feu quand ils font secs, sans se mettre en peine si ce sont des bois propres à bâtir ou non, ou si le tems est propre pour les abbatre et les conserver; mais ceux qui ont du bon sens et de l'économie aiment mieux n'aller pas si vite, et conserer tous les arbres qui sont bons à faire des planches, du cartelage, des poutres et autres bois de charpente, ce qui est un profit très-considerable, sur tout à present que les bois à bâtir deviennent très rares, et par conséquent très-chers. Il font donc attendre le déclin de la lune pour abbatre les arbres qui sont bons à quelque chose; les couper par tronses de la longueur qu'on juge à propos, les ranger les uns sur les autres et y faire un petit toit pour les défendre de la pluye, jusqu'à ce qu'on ait le loisir de les travailler. Après cela on amasse en plusieurs monceaux les branchages et les bois inutiles que l'on veut brûler: sur quoi il faut observer d'y mettre toûjours le feu sous le vent, c'est-à-dire,

du côté opposé du vent, après avoir fait une trace ou chemin bien net pour separer le terrain que l'on veut brûler, de celui qu'on veut conserver, et cela pour deux raison. La première, afin d'être toûjours maître du feu, et empêcher quand on le juge à propos qu'il n'aille trop loin, ce qu'on ne pourroit pas faire si le fent chassser avec trop de violence, et embraser les endroits qu'on veut conserver. La seconde, parce que le feu ne passant pas avec tant de rapidité, et comme en courant sur les endroits que l'on veut brûler, il a plus de tems pour consumer les bois abbatus et leurs fouches.'

18 Du Tertre, *Histoire générale des Antilles*, vol. 2, p. 451: 'Celles des officiers et des riches habitants ne sont pour la plupart qu'une charpente revêtue de planches, avec un étage au-dessus de la salle, dont le plancher est d'ais ou de brique; elles sont couvertes de tuiles [...]. Les autres ne sont couvertes que de bardeaux de bois, en guise de tuiles [...]. Les maisons des simples habitants ne sont encore palissadées que de roseaux, particulièrement là où on ne craint pas les incursions des sauvages [...] celles des plus pauvres sont couvertes de feuilles de cannes, de roseaux, de latanier et de palmiste; celles-là sont incomparablement plus agréables que nos chaumines de France.'

19 Ibid., p. 451ff.

20 Médéric Louis Elle Moreau de Saint-Méry, *Description topographique, physique, civile, politique et historique de la partie française de Saint-Domingue; avec des observations générales sur la population, accompagnées de détails les plus propres à faire connaître l'état de la colonie à l'époque du 18 octobre 1789* (Philadelphia, PA: Chez l'auteur, 1797–8), vol. 1, p. 11.

21 Christophe Charlery, 'Maisons de maître et habitations coloniales dans les anciens territoires français de l'Amérique tropicale', *In Situ. Revue des patrimoines* 5 (2004), http://insitu.revues.org/2362, retrieved 9 May 2017.

22 Labat, *Nouveau Voyage*, vol. 1, pp. 441–2: 'Mes charpentiers se trouverent en état de monter l'agrandissement de ma maison qui se trouva ainsi de trente-deux pieds de long sur seize pieds de large. La salle que l'on trouvoit en entrant avoit seize pieds en quarré. Les deux portes opposées répondoient à celle de la cour et à l'allée du milieu de mon jardin. La porte qui entroit de la salle dans ma chambre étoit à main gauche, elle avoit la meme grandeur que la salle, mais j'y avois fait un retranchement de cinq pieds de large sur toute la longueur qui me servoit à serrer mes provisions. J'avois ménage dans ce même espace l'escalier pour monter au galletas qui étoit assez commode pour y placer plusieurs hamacs; c'étoit la chamber de mon pensionnaire, où je me retirois aussi quand je donnois la mienne à quelque étranger. Je fis faire un perron de pierre de taille avec trois marches devant la porte de la salle, le reste du terrain alloit en pente douce pour donner lieu aux eaux de s'écouler.'

23 Du Tertre, *Histoire générale des Antilles*, vol. 2, p. 517.

24 Sen. epist. 90.10, transl. Richard Mott Gummere (Cambridge, MA: Harvard University Press, 1920).

25 Du Tertre, *Histoire générale des Antilles*, vol. 2, pp. 517–18: 'Elles [the African huts] n'ont guére plus de neuf à dix pieds de longueur sur six de large, et dix ou douze de haut; ells sont composées de quarter fourches qui en font les quarter coins, et de deux autres plus eslevées qui appuyent la couverture qui n'est que de roseaux, que la pluspart font descendre jusqu'à un pied de terre. Ceux qui la tiennent plus hautes, la pallissadent avec de gros pieux qui se touchent les uns les autres, sans se server de roseaux comme les François, qui sont bien aises d'avoir de l'air: si bien que leurs cases sont closes comme une boëte, de peur que le vent n'y entre, ce qu'ils font avec beaucoup de raison, parce que n'y estant Presque jamais que la nuit, comme ces nuits sont extrémement froides, ils seroient trop incommodez du vent, et du grand air, ainsi le iour n'y entre que par la porte qui est de cinq pieds de haut.'

26 See Debien, *Les Grand'cases*, p. 16.

27 E. Hayot, 'Les gens de couleur libres du Fort-Royal, 1679–1823', *Revue française d'histoire d'outre-mer* 56:202/203 (1969), 22–3. For the period from 1700 to 1823 Hayot lists for Fort Royal 24 African carpenters ('nègres'), 5 of them maîtres, 20

câpres (2 maîtres), 100 mulattos (11 maîtres), 38 métifs (6 maîtres), and 1 mamluk who was the offspring of a white man and a female métis (*métive*) (p. 24). Since 1723 he found that 58 free black people became masons, 26 became maîtres, and 6 contractors (*entrepreneurs*) (p. 28). He counts 54 free black cabinetmakers since 1710, 16 free Africans (7 maîtres), 1 câpre who was also a *maître*, 18 mulattos (4 maîtres), 5 métifs (3 maîtres), and 1 carteron or quarteron (p. 31).

28 See Charlery, 'Maisons de maître et habitations coloniales', who mentions some examples of such treatises circulating in Saint-Domingue, mostly concerning the ideal *villa rustica* that found adaptation in the colonies, by Louis Liger, *Œconomie générale de la campagne, ou La nouvelle maison rustique*, 2 vols (Paris: Charles de Sercy, 1700); Augustin Charles Daviler, *Cours d'architecture, qui comprend les ordres de Vignole* (Paris: Nicolas Langlois, 1691); Jean Antoine de Brûletout de Préfontaine, *Maison rustique, à l'usage de la partie de la France équinoxiale connue sous le nom de Cayenne* (Paris: J. B. Bauche, 1763); see also Emilie d'Orgeix and Céline Frémaux, 'La petite maison dans les abattis ou l'art de rédiger aux bois par Jean Antoine de Brûletout, chevalier de Préfontaine dans son habitation de la France équinoxiale (1754–1763)', *In Situ. Revue des patrimoines* 21 (2013), http://insitu.revues.org/10338, retrieved 23 May 2017.

29 Labat, *Nouveau Voyage*, vol. 3, pp. 417–38.

30 Ibid., p. 428.

31 Ibid., p. 431: 'Il est bon de distinguer toûjours les Negres ouvriers des autres, soit en leur donnant plus de viande, soit en leur faisant quelque gratification. Rien ne les anime davantage à chercher l'occasion d'apprendre un metier. Tel qu'il puisse être, il est toûjours d'une grande utilité pour une maison. Les profits que font les ouvriers, les attachment à leur maîtres, et leur donnent le moyen d'entretenir leurs familles avec quelque sorte d'éclat, et le plaisir d'être au-dessus des autres contente extrêmement la vanité dont ils sont très-bien pourvûs. J'en ai vû qui étoient si fiers d'être Maçons ou Menuisiers, qu'ils affectoient d'aller à l'Eglise avec leur regle et leur tablier.'

32 Ibid., p. 435: 'Il y a bien des habitans qui se servent plutôt d'un commandeur negre que d'un blanc. Sans entrer dans les raisons d'économie, je croi qu'ils font fort bien, et je m'en suis toûjours bien trouvé. Il faut choisir pour cet employ, un Negre fidele, sage, qui entende bien le travail, qui soit affectionné, qui sache se faire obéïr, et bien executer les orders qu'il reçoit; ce dernier point est aisé à trouver: car il n'y a point de gens au monde qui commandent avec plus d'empire, et qui se fassent mieux obéïr que les Negres. C'est au Maître à veiller sur ses autres qualitez.'

33 CAOM COL C8B 2: Guitaut/de Goimpy: Mémoire pour le Roy, 4 October 1691: 'il n'est pas possible d'en faire une estimation juste comme ceux qui sont employé a pescher la roche a chaue, Ils font plus de travail suivant que la mer est plus belle, les autres qui vont couper le bois porcuir cette chaue, a mesure qu'ils trouvent des bois plus moux ils en troncent et abbattent d'avantage et puis l'esloignement de l'embarquement avance ou differe le chargment des barques, qui les vont chercher, ces bastimens aussy sont quelques fois retenus par les mauvais temps, qui les empeschent de naviguer et dont l'entretien soit pour les voiles agrez et corps du bastiment, soit pour la solde et nourriture des equipages cauze des dépenses considerables.'

34 See CAOM COL C8B 22: La Varenne to the Council, 8 April 1717, fol. 96.

35 CAOM COL C8A 44: Houel to Maurepas, 1 November 1733: 'J'ay l'Honneur de vous informer que je suis party de la basse terre de la Guadeloupe le 23 octobre et que je suis arrivé icy le 26 du même mois après avoir esté obligé de faire cesser de travailler les maçons que j'avois employé à la construction du demy Bastion de la teste de la nouvelle fortification fautte de chaux, de sable, et de roches ayant employe generalement tous les matteriaux que M. Dupayet avoit fait ramaser pendant les six mois que j'y ay esté occupé. J'ay seulement faut continuer les tailleurs de pierres qui suivent les modeles de coupe que je leur ay donné avant de partir de cette isle pour me rendre en celle cy ou les travaux m'appelloient.'

36 Ibid.: 'J'ay laissé aussy les instructions necessaires pour faire l'amas des roches et de sable pour mettre le demy bastion a sa perfection avec ordre de bruler le peu de

roches a chaux qui restoit en attendant que les batteaux en apportassent d'autres, n'ayant encore eu que trois batteaux qui ayent satisfait de trente ou environ qu'il y a il est necessaire qu'on tienne la main a ce qu'ils fournissent regulierrement tous les ans, car si ces corvées sont negligées comme elles l'on esté depuis 1725 il est certain que ces travaux languiront considerablement, ce qui ne seroit point arrivé, si depuis ce temps en y avoit esté exact, au contraire, ces amas se seroient trouvez inseniblement faits et il n'auroit pas falû assurement trois années pour mettre cette fortification a sa perfection, ce qui est aisé à prouver a proportion de ceque j'y ay fait faire, ne s'agissant dans ces occasions que d'avoir l'œil a employer apropos les ouvriers et les manœuvriers que je ne quitte pas de vue du matin du soir.'

37 CAOM COL C8B 2: Guitaut/Dumaitz de Goimpy: Mémoire pour le Roy, 4 October 1691: 'il y a eu deux barques d'acheptez des fonds destinez pour Saint-Christophle et des avances faits aux entrepreneurs de la chaue jusques a la concurrance des quelles ils n'avoient pas fait de fourniture.'

38 Labat, *Nouveau Voyage*, vol. 1, p. 195.

39 Ibid., p. 196.

40 Hayot, 'Les gens de couleur', p. 38.

41 Du Tertre, *Histoire générale des Antilles*, vol. 2, p. 77.

42 Ibid., p. 79.

43 Ibid., p. 80.

44 Ibid., p. 81.

45 See Labat, *Nouveau Voyage*, vol. 3, pp. 3–15.

46 Ibid., p. 9.

47 CAOM COL C8A 2, fol. 415: Payen: Projets de travaux pour le Fort Royal, 30 November 1680: 'Le prix de chaque toise cube de massonnerie que l'on payoit l'année derniere à soixante livres, m'a paru sy execif que je me suis appliqué à le diminuer ayant examiné les matteriaux quy sont fort esloignés dud. fort, mais comme nous tirerons quantité de pierre et caillons de l'escarpement des ouvrage que je propose, ladit massonnerie se pourra faire a trente livres la toise cube pour n'en que nous soyons ay des de la flutte qui est arrivée depouis peu de Rochefort pour voiturer les bois et pierre a chaux qu'il faut apporter de Sr. Allouzie et de led. travailler.'

48 CAOM 08DFC/27, No. 204: Rochemore to Choiseul, 14 July 1765.

49 Ibid., No. 272: Le Beuf: Etat des fers, cloux, etc. etc. necessaires au service des fortifications, 15 August 1769.

50 Du Tertre, *Histoire générale des Antilles*, vol. 2, p. 451.

51 CAOM COL C8B 6, No. 4: La Roulais: Décision du Conseil de Marine relative aux états de matériel demandés, 5 January 1720.

52 Ibid.: 'Il faut examiner avec Mr. de la Roulaye ce qui se peut prendre a Paris de toutes ledites fournitures, et en faire marché donner a Rochefort les ordres necessaire pour ce qui devra este apris dans ce Port.'

6

Community and segregation in Louisbourg: An 'ideal' colonial city in Atlantic Canada

In contrast to maintaining that the French empire was a result of the action of a pluralist and heterogeneous colonial society the case of Louisbourg conveys that an ideal colonial city in the sense of representing France overseas was achieved to some extent, too. The well-studied history of this settlement is but an exception to the rule of French empire building. Colonial administrators pursued segregation and the installment of a national culture that remained predominantly white and excluded non-European groups that lived in the local vicinity. As in the case of Senegambia, the buildings constructed in Louisbourg also served an affective purpose and thus contributed to the establishment of an emotional community amongst the French population.

Style was an important element of French imperial symbolism. The spires of Louisbourg still loom into the sky of today's Cape Breton Island, formerly known as the Île Royale, off the mainland of Canada.[1] Since the Canadian government decided to 'symbolically reconstruct' the early modern fortress in the 1960s it has become a *lieu de mémoire* of the French heritage of the country. It is, of course, slightly misleading to speak of spires in plural, since only one has been rebuilt, or to give the impression that these are, by today's measure, particularly high. However, even though the historic buildings of the tourist site of Louisbourg cover only a fraction of the former extent of the fortified town (*ville fortifiée*), they still give the visitor a glimpse of early modern colonial life in Atlantic Canada.[2]

The largest reconstructed structure is the King's Bastion Barracks, crowned in the centre by the imposing bell tower with a large clock. It gives an idea of the dimensions and aesthetic appearance of colonial architecture in Louisbourg.

Claude-Etienne Verrier, the son of its architect and the town's chief engineer, Etienne Verrier, exaggerated its size in his coloured view of

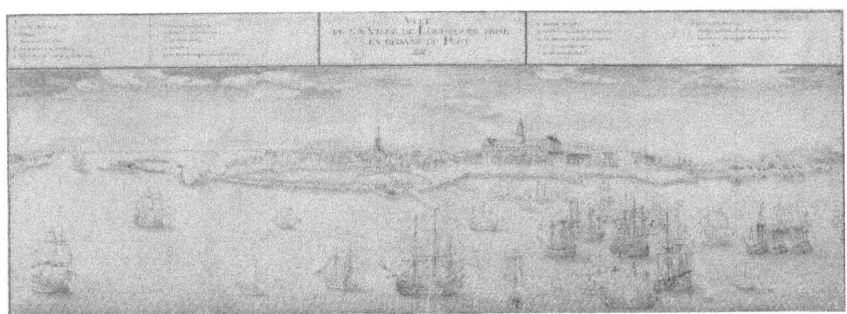

39 [Claude-Etienne] Verrier: Veue de la ville de Louisbourg prise dedans du port, 1731 (BNF Paris, Département des Cartes et plans, GE C-5019 (RES)).

Louisbourg from 1731 (see Figure 39). Here, the bell tower appears with the spire of the hospital, forming a skyline that became a well-known landmark of the North American coast. The sketch by the young Verrier highlights a lot of the detail of these two structures and a number of ornamental devices, such as the octagonal form of the hospital's tower, balanced above the roof atop its slim pedestal (see Figure 40). On top stood a weathercock that was complemented by the weathervane of the barracks' clock tower. This prominent tower was more compact and rested on a turret of masonry. A design from 1733 and its current reconstructed state both reveal a square structure crowned by a Fleur-de-Lys (see Figure 41).

Some historians describe the fortified city of Louisbourg as the crown jewel of the French empire in the eighteenth century – 'the key to [French] possessions in North America' (Voltaire).[3] From its foundation in 1713 it was a commercial centre of the Atlantic fishing industry and occupied an important strategic position for the military defence of New France. Also, many ships to and from France, the West Indies, Canada, and the British colonies in New England and Nova Scotia dropped anchor for resupply and to load or unload commodities.

The enormous efforts made by the administration in Versailles to maintain the stronghold for military purposes serves as proof of its imperial ambition, which is most often attributed to naval secretary Jean-Frédéric Phélypeaux de Maurepas. His central mission of co-ordinating the navy and colonial commerce, in the sense of directing the vessels of the *Royale* to protect merchant convoys as its primary purpose (*protection du commerce*), found perhaps its clearest expression in the committing of large sums from the navy's budget to the improvement of Louisbourg.[4] No other colony received more money. According to Johnston, its expenses counted for 10 per cent of the combined

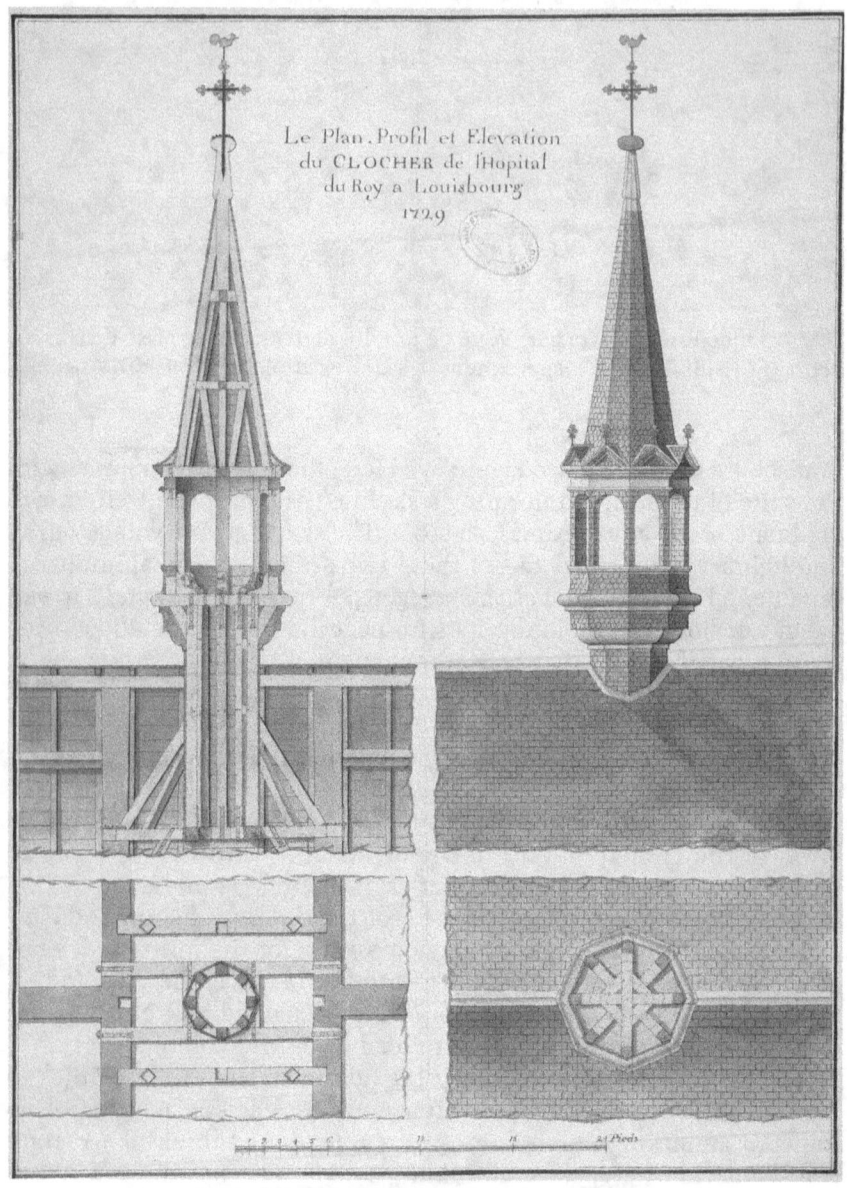

40 Le Plan, Profil et Elévation du Clocher de l'Hopital du Roy à Louisbourg, 1729 (CAOM C11B 39/110bs).

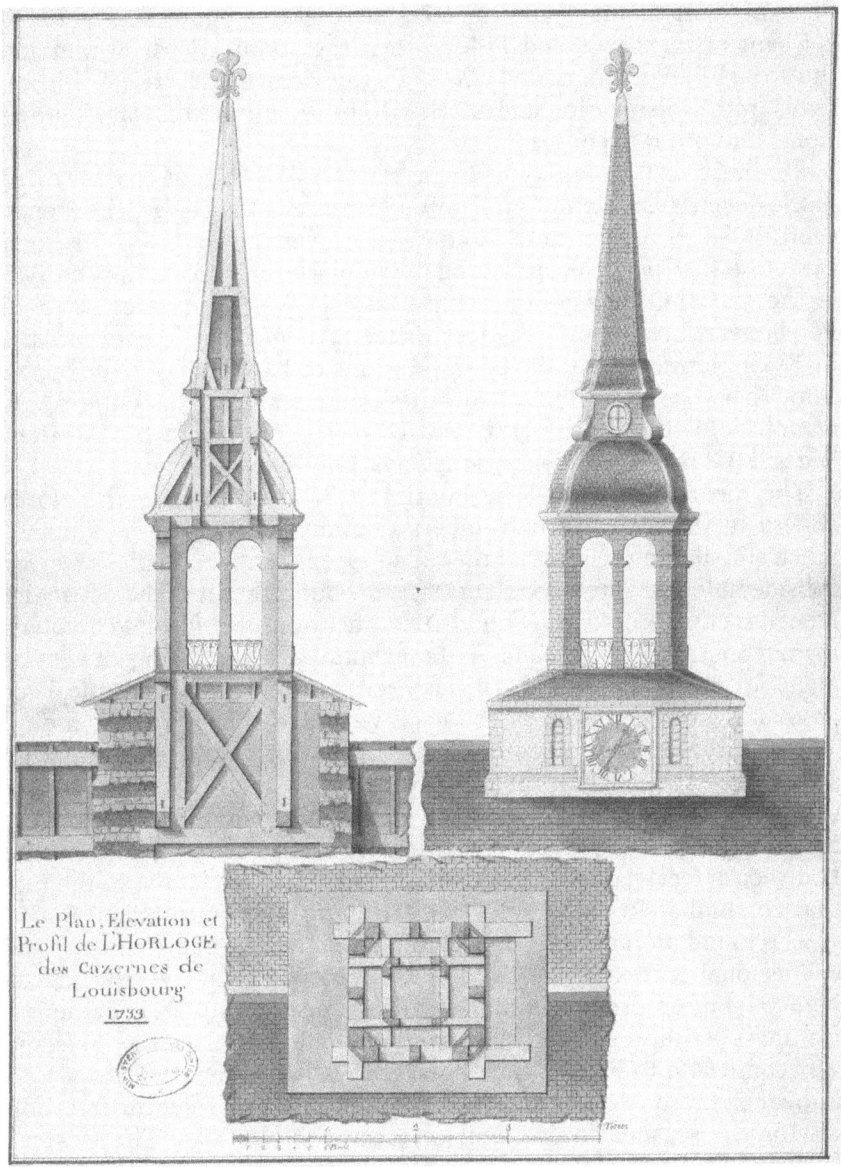

41   Anonymous: Le plan, elevation et profil de l'Horloge des Cazernes de
Louisbourg, 1733 (CAOM Col C11B 39, 109).

household for the navy and the colonies, and this figure rose to 20–30 per cent in the 1730s and 1740s, when the conflict with Britain and the Royal Navy intensified. In total, the Crown invested 4 million livres in the fortification and another 16 million livres on wages, provisions, and other expenses.[5]

Similarly to Pondichéry in India, the fate of Louisbourg was sealed in the Seven Years' War. On 27 July 1758, after a long siege, the French capitulated to British forces and handed over the town for a second time, the first occupation having taken place in 1745 during the War of the Austrian Succession (King George's War). But in contrast to Pondichéry, Louisbourg was lost to the British in the Treaty of Paris in 1763, along with all of the possessions of New France. The 'chess game' between the French and British empires was thus entering its last phase, with France being reduced to its colonies in the Antilles, Senegal (Gorée), the Mascarene Islands, and India.[6]

The fortified town of Louisbourg is a good example of the North American variant of French imperial building activity. It was not so much singular in the sense of the money spent – other colonies received considerable sums, too, as we have seen – but rather it is the particular French style of architecture and the distinct mixture of people involved in its construction that makes Louisbourg an interesting case. The plans for the hospital and barracks sent to Versailles were ideal for producing an impression of the local visibility of the French empire. The apocryphal sentence, attributed to Louis XV, that he must be seeing the spires of Louisbourg in view of the high costs of a project is of anecdotal value in conveying an idea of the metropolitan perception, or better non-perception, of what happened in the colony.[7] The attention Louisbourg received temporarily might be a small exception to the rule that colonial affairs were rarely a subject of importance at court.

The foundation of Louisbourg, in fact, has to be placed in the local and regional context of the North Atlantic. The French style introduced here was, as in other colonial localities, not merely an ornamental feature superimposed on diverse building traditions. Cape Breton Island was connected to France via a passage that had come to be very well known to Breton and Basque fishermen during many centuries of following the seasonal movement of oceanic fish stocks. The western port cities of Saint-Malo, Brest, Rochefort, Bordeaux, and Bayonne were culturally and socially much more closely connected than Versailles. Most of the populations of Louisbourg and the Île Royale, which peaked at 4,174 and 5,845 inhabitants respectively in 1752, came from the Atlantic coast of France and were mostly skilled artisans and professionals in fishery and whaling.[8]

Engineers could rely on the available craftsmanship on Île Royale during the development of the fortified town, and this enabled more elaborate building projects to be undertaken in Louisbourg than in other colonial places.

The refinement of the carpentry for the wooden structures of the turrets on the barracks and the hospital has already been mentioned at the start of this chapter. Likewise, the architecture of the three main gates of the city walls required highly skilled masons and stone dressers. Verrier designed a tasteful gate for the Dauphin half bastion in the north-western section of the fortification (see Figures 42 and 43). Not large in overall size, the arch carrying the royal coat of arms was flanked by two panelled buttresses with stylized armoury figures on top. Facing the city, a plaque with an inscription praising Louis XV was fitted on the other side of the coat of arms over the arch.

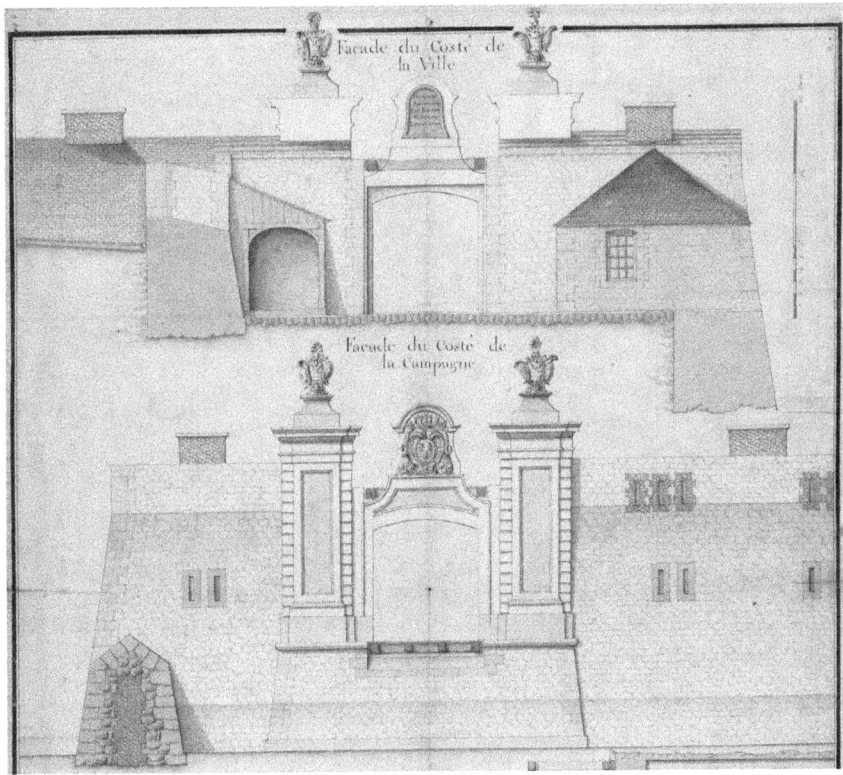

42 Anonymous: Plan de la Porte Dauphine de Louisbourg, 1733 (CAOM Col C11B 39, 41).

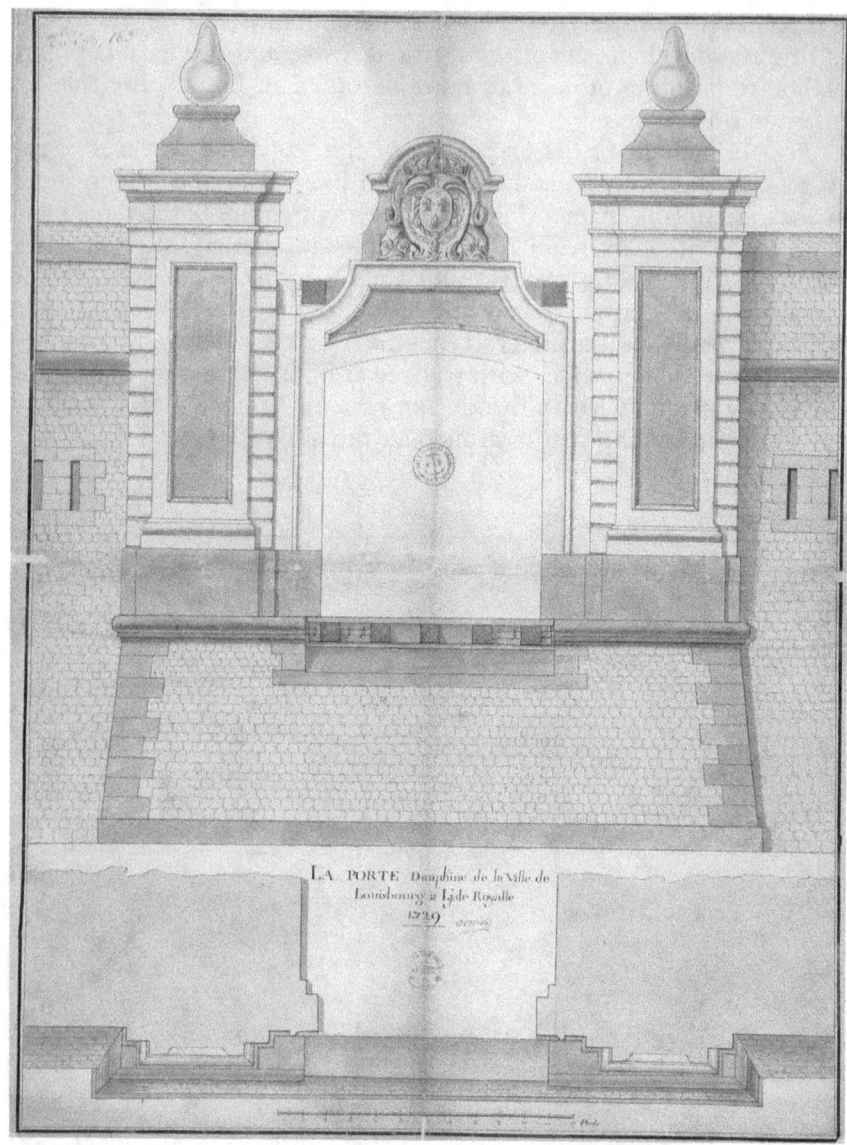

43 Etienne Verrier: La porte Dauphine de la ville de Louisbourg à l'isle Royale, 1729 (CAOM 03DFC/163B).

Verrier's initial plan for the Porte de Maurepas, part of the later fortifications added to the east side, had a classical pointed pediment, but this was replaced by a rounded pediment in a second design dated 1741 (see Figures 44, 45, and 46). The engineer conceived an even

[126]

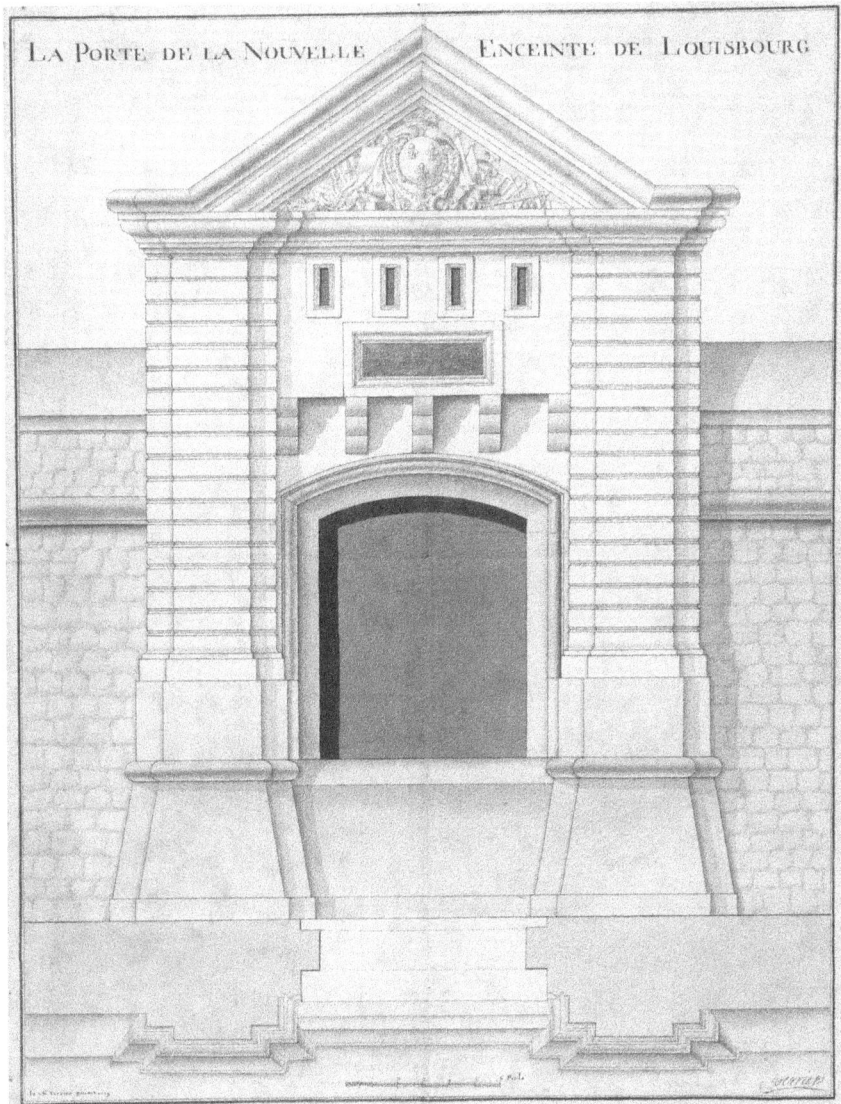

LA PORTE DE LA NOUVELLE   ENCEINTE DE LOUISBOURG

44 Etienne Verrier: La Porte de la Nouvelle Enceinte de Louisbourg, 1739 (CAOM Col C11B 39, 44).

more imperial look for this gate, adding a royal coat of arms garnished with a multitude of naval and army ordnance and four Fleurs-de-Lys on top of the roof. The city façade kept the pointed pediment, and below that the inscription '*Porte de Maurepas*' was to be carved in

[127]

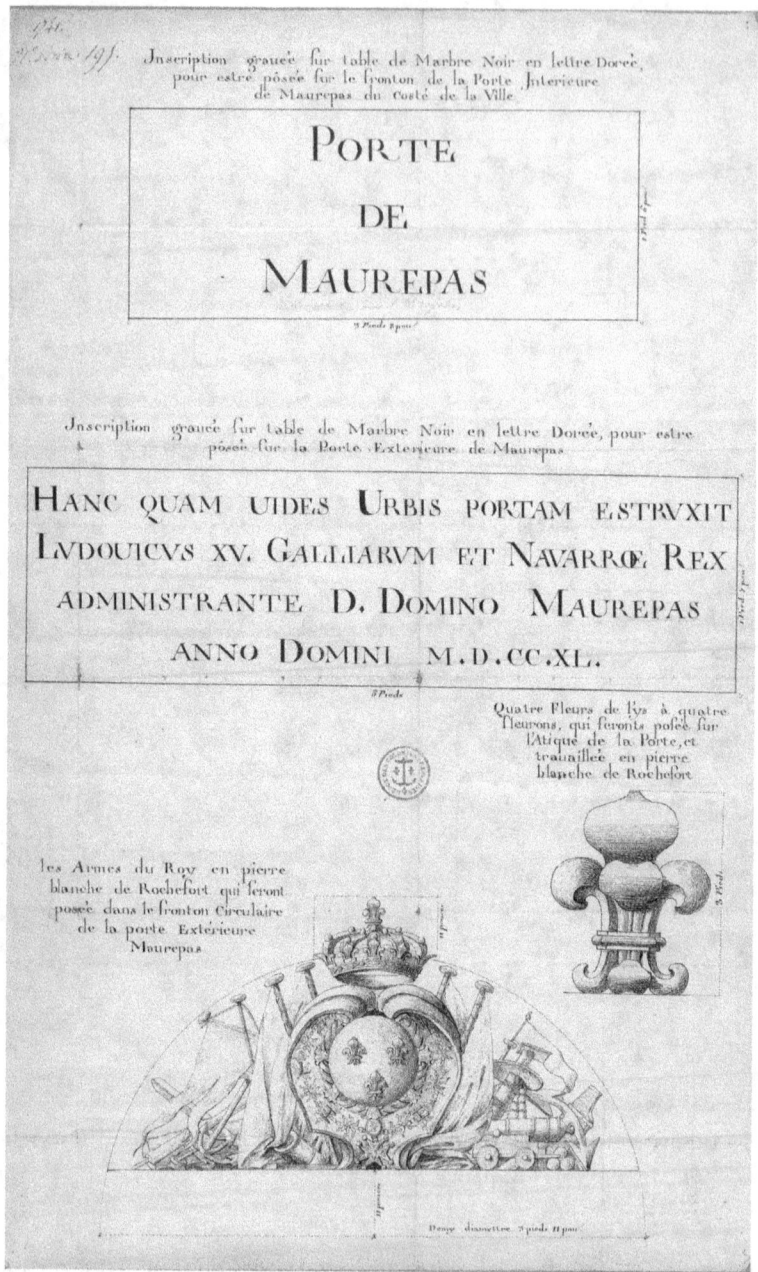

45 Etienne Verrier: Les différents profils de la Nouvelle Enceinte de Louisbourg, où on a représenté [...] Porte de Maurepas, 1741 (CAOM 03DFC/194B).

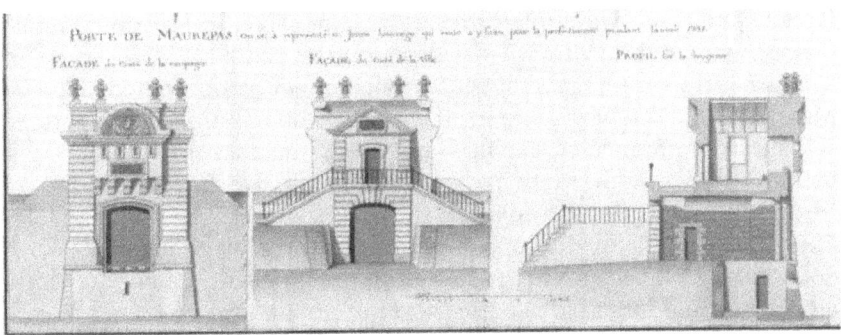

46 Etienne Verrier: Porte de Maurepas, 1741, Detail (CAOM 03DFC/195C).

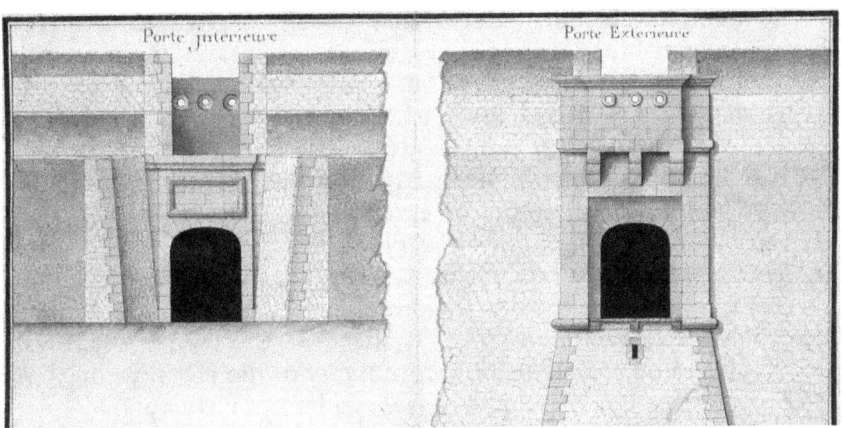

47 Anonymous: Plan, profil et elevation de la Porte de la Reine,
dans une des Courtines de l'Enceinte de la ville de Louisbourg, 1733, Detail
(CAOM Col C11B 39, 46).

black marble. The Porte de la Reine, on the other hand, seemed plain
compared to the other two, yet still monumental in its defensive function;
only three bullseyes in the parapet gave it a characteristic look (see
Figure 47).

Verrier, who was responsible for most of the aesthetic appearance
of Louisbourg, relied not only on French expertise, but also on building
materials imported from France.

His detailed estimates for construction projects show the use of
stone, bricks, and wood mostly from the islands, but sometimes also
from France. While the chimney of the oven for the bakery, for example,
was to be built of oak timber, and the oven itself of large stones made
in the country, for the chimney cowl, stones were brought over from

France, and these were five times more expensive. In addition, French timber was used for mortar platforms on the parapets of the wall.[9]

The contractor for the above-mentioned minor works was one Bernard Muiron, while the workforce itself was made up of European indentured servants from France or, during wartime, prisoners from other nations. In 1690, for example, the engineer responsible for the fortress of Port Royale in Acadie (later Annapolis in Nova Scotia) proposed using forty English prisoners for the construction of covered ways and palisades.[10] Soldiers also played an important part in the fortification effort, eventually forming a large part of the working population of French North America.[11] Indigenous slaves were practically absent and those few African slaves on Île Royale were mostly employed in domestic service, although some worked as stone masons or ferrymen.[12]

The architecture contributed in many senses to the feeling of community in Louisbourg. Safely confined, the inhabitants of the fortified town could feel protected against foreign invasion, but also against any intrusions by their indigenous neighbours, the tribes of the Mi'kmaq nation. When Louisbourg returned to France after the Treaty of Aix-la-Chapelle in 1748, the French garrison made its entry through the Porte de la Reine.[13] There was an emotional aspect to this re-entry, as the most beautiful gate, Porte Dauphine, had been destroyed. Louis Franquet, the new engineer in charge of repairing the fortifications, did his best to restore the town to its former glory, but his plans remained mostly on paper and between 1754 and 1758 the French spent only 480,000 livres on repairs and improvements to the fortifications.[14]

The practice of displaying the royal symbols of power and of aristocratic lifestyle, however, continued under the new commander, Jean-Louis de Raymond. In 1751, he arranged banquets and balls, and held ceremonies for the King's victories or to celebrate important events. For example, the birth of the Duc de Bourgogne was marked with a Te Deum and artillery salutes. What must have been a wondrous sight for the people of Louisbourg unfolded before them on a Sunday evening, 28 May:

> Raymond gave dinner to the staff, the engineers, the officers of artillery, and to the other principal officers, to the Conseil Superieur, the Baillage, the Admiralty, and to the ladies of the place. He had two tables with 50 covers, served in four courses, with as much lavishness as elegance. They drank in turn freely every kind of wine of the best brands, to the health of the King, Queen, the Dauphin, Mme. la Dauphine, M. le duc de Bourgogne, and to the Royal Princesses. Many guns were fired, and the band added to the festivities. About 6 o'clock, after leaving the table, they repaired to the King's chapel to hear vespers. At the close of the service, the Te Deum was sung to the accompaniment of all the artillery of the town and of the ships. Then they went in a procession, as is the

custom in the colonies, to the Esplanade of the Maurepas gate. The governor there lit a bonfire, which he had prepared; the troops of the garrison, drawn up on the ramparts and the covered way, fired with the greatest exactness three volleys of musketry, and the artillery did the same. After this ceremony, the Governor distributed several barrels of his own wine to the troops and to the public. The 'Vive le Roi' was so frequently repeated, that no one could doubt that the hearts of the townspeople, the troops, and the country folk, which the festival had attracted, were truly French. He had given such good orders to establish continual patrols under the command of officers that no disorder was committed. About 9 in the evening, the governor and all his guests went to watch the fireworks and a great number of rockets, which he had prepared, and which were very well done. On his return home, the ball was opened, and lasted till dawn; all kinds of refreshments, and in abundance, were handed round. His house was illuminated with lanterns placed all around the windows, looking onto the rue Royale and the rue Toulouse. Three porticoes, with four pyramids, adorned by triple lanterns and wreaths of flowers, rare for such a cold climate, were erected opposite the rue Royale. At the opposite corner, where the two roads cross, two other pyramids were also illuminated; and on the front of the three porticoes were painted the arms of the King, the Dauphin, and the Duc de Bourgogne. At the end of the same street, opposite the three porticoes, were also represented, by means of lanterns, three large Fleurs-de-Lys and a 'Vive le Roi', very visibly placed on a banner above. Between these two principal illuminations was situated the large gate of the Government House, which was also adorned by triple lanterns on its columns and cornices; above was the King's portrait. All round the courtyard there were also fire pots and triple lanterns, as high as the retaining wall of the garden. These illuminations were charming in their effect and lasted till the end of the ball; all the houses in the town were also lit up, as well as the frigate *la Fidele*. The government house being too small to accommodate all the distinguished members of the colony, M. le Comte Raymond gave a big dinner the next day to the clergy and the Sunday following to several ladies, officers, and others who had not attended the first celebration. It can be said that the Governor held nothing back for these festivities and that he gave on that happy occasion very evident proof of a rare generosity.[15]

The fortified town of Louisbourg had all the features of a European town. Predominantly occupied by Frenchmen and Basques, the cityscape and the geometric outline of the fortifications had the appearance of one of the newly planned Baroque cities in France or Germany (see Figures 48 and 49). The administration established a political culture in which the status not only of the elite of society but also of large parts of the population mirrored the set-up of the metropolitan courtly system. Louisbourg thus appears to have been a tightly knit emotional

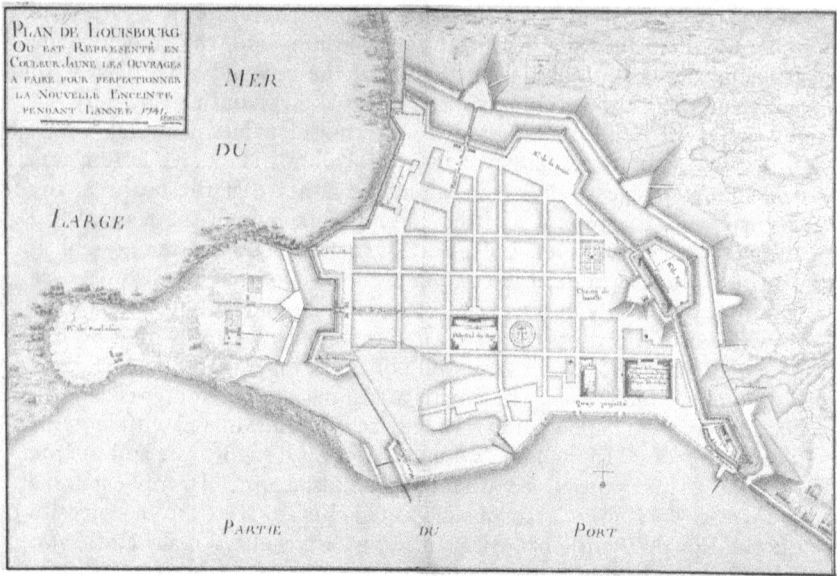

48 Etienne Verrier: Plan de Louisbourg, où est representé en couleur jaune les ouvrages à faire pour perfectionner la nouvelle enceinte pendant l'année 1741 (CAOM 03DFC196B).

community. The belligerent atmosphere during the 1740s and 1750s between the two rising imperial powers of France and Britain contributed to a common identity. This feeling, of course, only intensified during the weeks and months of the two sieges, through the experiences of forced removal, exile, and return.[16]

The spatial order of Louisbourg, however, revealed two exclusionary practices of this local expression of empire. First, the native population does not appear in the context of the political, social, and cultural life of the city; neither were Mi'kmaqs or members of other indigenous nations included in the building effort of the large colonial projects of Louisbourg. This is a remarkable exception in light of the different expressions of local French imperial identity discussed in this book. Without elaborating on this aspect in detail, the exclusion of an indigenous workforce, expertise, and knowledge from the execution of large construction projects was typical for New France. In Quebec, Montreal, Trois-Rivières, and Port Royale the French relied only in the very early stages on indigenous collaboration to establish their settlements. Historians described Quebec in the seventeenth century as a traditional French town that relied heavily on the division of society into a colonial elite, craftsmen, artisans, and the indentured servants (*engagés*).[17] This

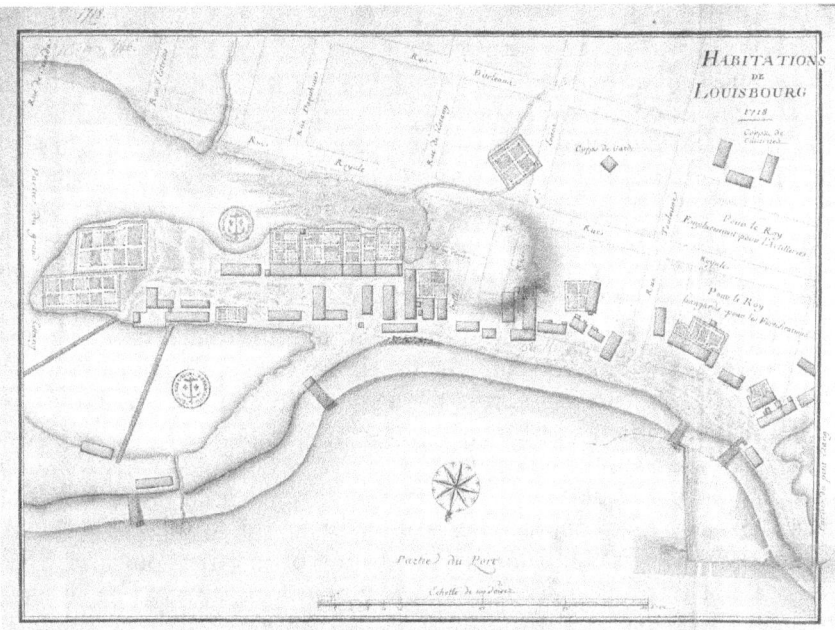

49   Jean-François du Vergery de Verville: Habitations de Louisbourg, 1718 (CAOM 03DFC146bs). This plan shows the discrepancy of the planned grid structure for the colonial city and the actual situation of buildings scattered against the geometrical order along the coastline.

feudal asymmetry provided a sufficient workforce of mainly Europeans to improve the Canadian and Acadian settlements and transform them into colonial cities.

In Louisbourg this spatial order of enclosure that separated French and indigenous inhabitants was only interrupted during wartime. Particularly during the Seven Years' War, the French as well as their British counterparts tried to establish ties and alliances with the aboriginal nations. In the summer of 1757, the coastal posts above and below Louisbourg became the temporary homes of native warriors and Acadian refugees living together with French infantry, troops of the Compagnies franches de la Marine, and local militia. Mi'kmaqs from across the Atlantic coastal region and warriors from other nations came to join the fight to protect the stronghold against the invasion. This indispensable military support consisted mostly in pioneering and guiding French regular troops through the country. In return the Mi'kmaqs received 60,000 livres worth of weapons, tobacco, and assorted merchandise.[18]

[133]

The alliance, however, did not influence the treaty negotiations in Paris, during which the land of the Mi'kmaq, the Île Royale, known to the indigenous nation as Mi'kma'ki, was turned over to the British to become Cape Breton Island without further consultation. Even if the Mi'kmaqs had adapted to the imposition of a spatial order whereby land was divided into territories along geometric lines, their appeals for recognition of their property rights over the land remained unheard. The words of the Mi'kmaq chief, for example, written to the British administration in Halifax in 1749 provide an example of opposition to the European conception of *dominium*, the right to possess property, through the proposition of a native 'natural' right to appropriate space according to their terms: 'This land belongs to me. I have come from it as certainly as the grass, it is the very place of my birth and of my dwelling, and this land belongs to me.'[19]

The second spatial consequence of fortifying a whole town according to a preconceived plan was the practice of territorial enclosure that was pursued in metropolitan France as well as in the Antilles. Louisbourg was only one of several later realizations of fortified outposts that delineated a territory that the French claimed as *imperium*. It is difficult to prove that maps highlighting fortified places were used in the 'imperial centre' of Versailles to visualize their imaginary North American empire. Maps, however, are more than 'a mute testimony of forgotten struggles', as Charles Balesi wrote about the demise of the French North American communities after the Louisiana Purchase.[20] The string of fortresses on Île Royale and in Nova Scotia, the Ohio Valley, and the Upper Country (*Pays d'en haut*) does not only represent the everyday struggle of explorers, settlers, and soldiers building fortresses. What the maps also represent is the continuing extension of space throughout the 'middle ground', through the setting up of one post after another, introducing the idea of territorial enclosure through a series of physical structures.[21]

The spatial regime of Louisbourg was thus an example not only of the segregation of the European and indigenous populations on Île Royale. By reintroducing the erection of the physical fence, palisade, or wall not just for military protection – also present in indigenous societies – but also to protect the culture of a Eurpoean community in the form of the *ville fortifiée*, the example of Louisbourg was able to serve as a model for larger territorial spaces. Tragically, it was exclusively the Europeans who carried out the territorial expansion project of empire building in North America. Not only were indigenous people not involved in this endeavour, its execution was also performed at their expense. Thus, the segregational practice of Louisbourg, and of other comparable cases in North America, led to the most extreme expression of an imperial identity in the whole of the French colonial realm.

Notes

1 General accounts are offered by Johnston, *Control and Order*; A. J. B. Johnston, *Endgame 1758: The Promise, the Glory, and the Despair of Louisbourg's First Decade* (Lincoln, NA/London: University of Nebraska Press, 2007); J. S. McLennan, *Louisbourg from Its Foundation to Its Fall* (1918; repr. Halifax: Book Room, 1983); for a comparative account, see John Robert McNeill, *Atlantic Empires of France and Spain: Louisbourg and Havana, 1700–1763* (Chapel Hill, NC: University of North Carolina Press, 1985).
2 For this aspect of everyday life in Louisbourg, see A. J. B. Johnston, *Life and Religion at Louisbourg, 1713–1758* (Montreal and Kingston/London: McGill-Queen's University Press, 1996).
3 See Johnston, *Endgame*, pp. 13–14; Voltaire, 'Précis du siècle de Louis XV', in *Œuvres historiques* (Paris: Gallimard, 1957), p. 1462.
4 See Michel Vergé-Franceschi, *La Marine française au XVIIIe siècle* (Paris: SEDES, 1996).
5 Johnston, *Endgame*, p. 16.
6 Ibid., pp. 6–9.
7 See ibid., p. 17.
8 B. A. Balcom, *The Cod Fishery of Isle Royale, 1713–58* (Ottawa: National Historic Parks and Site Branch Parks Canada, 1984), pp. 4, 11–19. Despite its distance the eastern and western coasts of the Northern Atlantic Ocean were so closely connected that even germs could travel to and fro like, for example, in 1758 when a typhus epidemic quickly spread from Louisbourg to Brest and across Brittany (see Johnston, *Endgame*, p. 139). See also Pritchard, *In Search of Empire*, p. 33; A. J. B. Johnston, 'The Fishermen of Eighteenth-Century Cape Breton', in Eric Krause, Carol Corbin, and William O'Shea (eds), *Aspects of Louisbourg: Essays on the History of an Eighteenth-Century French Community in North America* (Sydney, Nova Scotia: University College of Cape Breton Press, 1995), pp. 198–208.
9 CAOM 03DFC/6, No. 200: Verrier: Toisé provisionnel des ouvreages qui ont été faits à l'Isle de l'entrée du Port, pour l'etablissement d'un Boulangerie, d'un Corps de garde pour les Soldats, dune prison et deux plattes formes de mortiers; par le Sr. Muiron Entrepreneur des fortiffications de Louisbourg pendant la presente année 1744, 30 October 1744, fol. 5r–v.
10 CAOM 03DFC/6, No. 69: Memoire Fortifications du Port Royal, 1690.
11 Leslie Choquette, *Frenchmen into Peasants: Modernity and Tradition in the Peopling of French Canada* (Cambridge, MA/London: Harvard University Press, 1997), pp. 13–14. In the early 1720s one-third of the entire population of Louisborg were soldiers (see A. J. B. Johnston, 'The People of Eighteenth-Century Louisbourg', *Nova Scotia Historical Review* 11:2 (1991), 75–83).
12 Kenneth Donavan, 'Slaves and Their Owners in Ile Royale, 1713–1760', *Acadiensis* 25:1 (1995), 4.
13 See Johnston, *Endgame*, p. 49.
14 Ibid., p. 50.
15 CAOM Col F3, Collection Moreau Saint-Méry, vol. 50, 3, pp. 420–23; trans. McLellan, *Louisbourg*, pp. 192–3.
16 See Johnston, *Endgame*, p. 47: 'It is possible that many returnees came back to Louisbourg with the beginnings of a common identity beyond that of being French-speaking and Roman Catholic.'
17 See Rémi Chénier, *Québec: A French Colonial Town in America, 1660 to 1690* (Ottawa: Minister of the Environment, 1991), p. 18; Marc Vallières, *Quebec City: A Brief History* (Quebec: Les Presses de l'Université Laval, 2011), p. 49.

18 Johnston, *Endgame*, p. 130. See also B. A. Balcom, 'Defending Unama'ki: Mi'kmaw Resistance in Cape Breton, 1745', *Nashwaak Review* 22–23 (2009), 447–92; for a general account of involvement of indigenous nations in war activities, see Fred Anderson, *Crucible of War: The Seven Years' War and the Fate of Empire in British North America, 1754–1766* (London: Vintage, 2001), esp. p. 848 for a list of 'Indian Peoples'.

19 Letter of 1748, quoted in Johnston, *Endgame*, p. 161.

20 Charles John Balesi, *The Time of the French in the Heart of North America: 1673–1818* (Chicago, IL: Alliance française, 1992), p. 326.

21 See the literature on the struggle of competing empires and nations in this region that itself proposes to consider the mutual dependency of indigenous and European actors in the process of empire and nation building: Kathleen DuVal, *The Native Ground: Indians and Colonists in the Heart of the Continent* (Philadelphia, PA: University of Pennsylvania Press, 2007); Gilles Havard, *Empire et métissages: Indiens et Français dans le Pays d'en Haut, 1660–1715* (Lille: Septentrion, 2003); White, *Middle Ground*.

7

Motley style: Affective buildings and emotional communities on Martinique, Guadeloupe, and Haiti

Early modern colonial empires lacked the supposed homogeneity of those of the modern era. The last case study returns to the Caribbean in order to illustrate how the techniques of empire building were not necessarily restricted to the use of European colonizers. Affective buildings and the resulting emotional communities are characteristic to early modern colonial culture, influenced government and social order, and could eventually be appropriated for different purposes, too. Thus the Haitian Revolution can serve as an event that marks the end of a period of trying to streamline the European effort to establish an imperial order. The successful revolt was not counterpoint to the tradition of combining the resources to pursue large building projects; in fact, it was based on it.

Affective buildings

Buildings, and large buildings in particular, might have affected contemporaries in different ways. Affective buildings did not necessarily awaken feelings of awe or appreciation, but may instead have given rise to feelings of alienation, rejection, or even disgust. Sources mostly inform us about these effects indirectly by providing insights into the reactions of others, but they can also directly reveal what impressions it was hoped some buildings would have on other people. There is, of course, a play of intended and unintended effects at work that sometimes makes it difficult to determine whether buildings were aesthetically appreciated or not. The aim of this chapter, however, is not to determine whether buildings were beautiful or not, but to establish that buildings were indeed designed to carry an emotional and symbolic message that eventually contributed to the emergence of a social formation in the sense of an emotional community.

The first representative building on Martinique, the house of governor Parquet in the Quartier du Carbet, was erected to impress the indigenous Carib people. Rochefort, ever enthusiastic about French accomplishments in the Antilles – a notion du Tertre never tired of criticizing as insincere and exaggerated – tells the story:

> The Indians, who had never seen a building of a similar outline, nor of a material that solid, examined it in the beginning with a profound astonishment, and later on tried to see if they could break it down with the force of their shoulders; they were obliged to avow that if all the houses were built like this the storm that is called Hurricane cannot damage them.[1]

The encounter demonstrates the need for the French to earn recognition for their functional building techniques, the aesthetic dimension obviously not triggering much excitement. At the beginning of colonization, representative architecture was rare in the French Antilles. Yet, while the large castles that were built in the first half of the seventeenth century may have left an impression on the people living in the vicinity, the level of building activity would rise, and with it would come the emergence of more elaborate styles.[2]

One can roughly distinguish between four separate styles. First there were the planned public buildings, such as seats of government, fortresses, ports, gates, or hospitals; second, the indigenous architecture that was extant in the seventeenth century, but which dissipated throughout the eighteenth century; third, African architecture with its characteristic techniques and ornaments; and fourth, something known as Creole architecture, a new style that emerged in the course of the formation of a prosperous slave economy in the Antilles.

This latter style developed further during the nineteenth century and is still visible on the islands today in a tradition that can traced in individual houses; it has already received some attention by scholars.[3] Carib architecture is no longer physically present with the exception of a small number of elements discoverable in newer buildings, such as the wattle panels made of reed or sugar cane tops (see Figure 50).

Certain African architectural features have also survived the early modern period in this way. Otherwise, it is just the large structures that remain.

The monumental nature of the first category of architecture can still be seen today in the island's landscape. Fort Royal, for example, remains a landmark of modern Fort-de-France on Martinique, Fort Saint-Charles on Guadeloupe survives, and even the city layout of modern Cayenne echoes the early modern fortification works. On the other hand, other

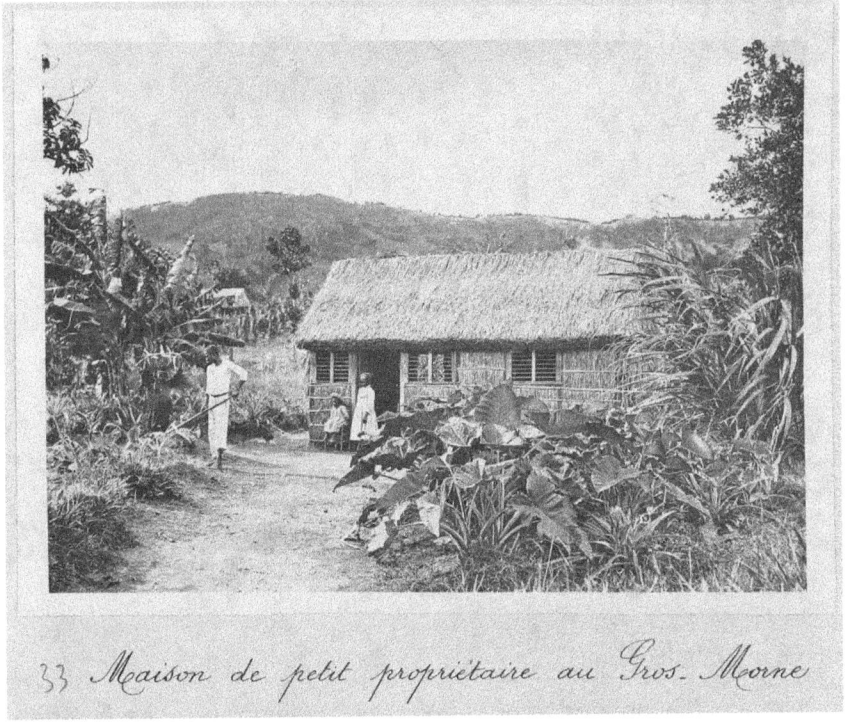

33 Maison de petit propriétaire au Gros-Morne

50   Camille Le Camus: Maison de petit propriétaire au Gros Morne,
1902–1907 (CAOM 8Fi73/33).

public works deteriorated and have been replaced by more recent
edifices.[4]

Inasmuch as it is possible to distinguish the four style categories,
however, they were often intertwined, and sometimes more than one
style was used for the same building. Examples of this motley style
are seen in the residential houses of free Afro-Caribbeans, which combine
the basic European structure with panels of wattled reed, a thatched
roof, and a floor of tamped earth. Modern photographs show these
workers' cottages (*cases à travailleurs*) as resembling the early modern
descriptions. While most of them were probably wooden, there were
also some built of stone, more an indication perhaps of the availability
of this resource within the local vicinity than of a higher living standard.
On a photograph taken by Reverend Jean-Baptiste Delawarde in 1935
we see an example of how Afro-Caribbean architecture was combined
with a Creole-style porch and ornamental pelmets (see Figure 51).

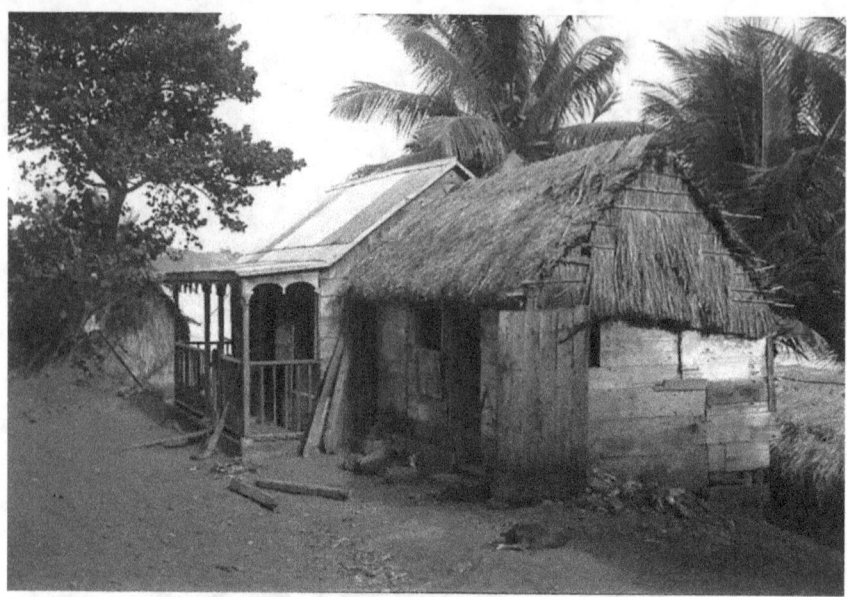

51 Jean-Baptiste Delawarde: Case de la Martinique, 1935 (AD Martinique, 31Fi00111).

The affective dimension of such cottages may not seem evident from the tranquillity these images convey. More than being calm they represent the confinement of the plantation, where the huts pointed to both the institution of slavery and an imagined space of marronage and freedom lying beyond this zone. Thus, the photograph can be interpreted as a memory of slavery, a 'symbolic anchor' to a past in which freedom was merely a dream. Aimé Césaire perhaps best articulates such a feeling in his magisterial poem *Cahier d'un retour au pays natal* from 1939:

> Au bout du petit matin, cette ville plate – étalée, trébuchée de son bon sens, inerte, essoufflée sous son fardeau géométrique de croix éternellement recommençante, indocile à son sort, muette, contrariée de toutes façons, incapable de croître selon le suc de cette terre, embarrassée, rognée, réduite, en rupture de faune et de flore. [...] Dans cette ville inerte, cette étrange foule qui ne s'entasse pas, ne se mêle pas: habile à découvrir le point de désencastration, de fuite, d'esquive.

> At the edge of dawn, this worn out town – laid out, collapsed and senseless, inert, winded under the square-edged burden of an eternally recommencing cross, not accepting its fate, silent, frustrated in every way, unable to grow by drawing sap from this earth, hampered, clipped, reduced, at odds with the natural world. [...] In this inert town, this strange crowd that

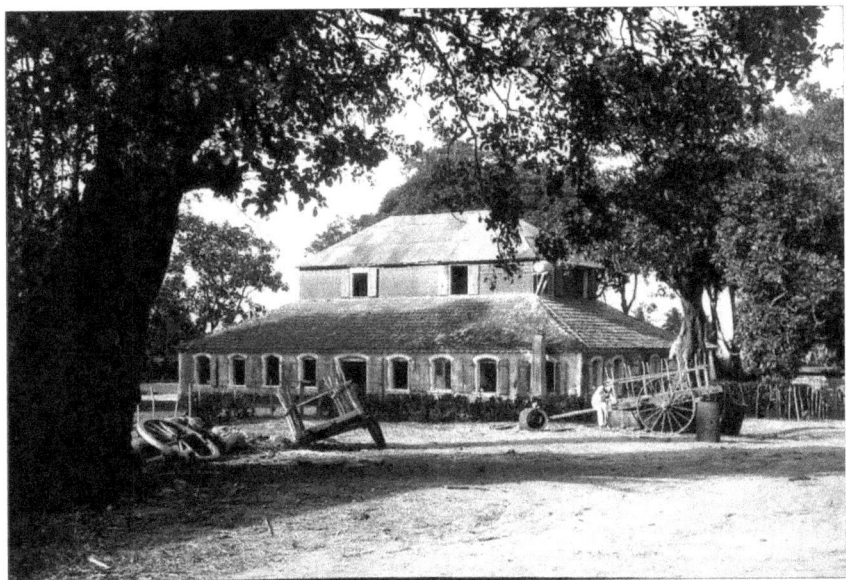

52 Jean-Baptiste Delawarde: Habitation de Pécoul, Basse-Pointe, 1937 (AD Martinique, 31Fi10). The house served as a model for the plantation mansion in the Antilles.

does not huddle together, does not mix; skilled at spotting the moment at which to slip away, to escape, to sidestep.[5]

The workers' cottage might have been the most common architectural feature of the Antilles. Planters' mansions and town houses, however, were more or less styled as *maisons creoles*, featuring elaborate structures, original ornamentation, and coloured façades. According to Crain, houses in Saint-Domingue, later Haiti, were painted in brighter colours than on Martinique. Here, lighter tones were chosen, while town houses were sometimes painted white.[6] Europeans derived their carpentry techniques from the shipyards, a fact that led to structures such as the master's mansion of the sugar plantation Pécoul in Basse-Pointe (see Figure 52). Its quadrilateral, red-tiled roof, and the galleries surrounding the upper storey that were added in a second phase of construction in the 1770s, are features characteristic of this Creole-colonial style.[7]

Architectural elements such as the gallery and tiled roofs conveyed the sort of refinement that distinguished the planters' mansions from their Afro-Caribbean neighbours. Besides their practical function of ventilating the walls against the humidity of the tropical climate, they are a symbol of Creole confidence.[8] Another interesting case is the house of the Sieur Banchereau in Carbet, the Carib village where the governor Parquet had his first residence. A plan of 1726 shows the façade of the Banchereau mansion with a gallery, two storeys, and a tiled roof. According to Charlery, it is the oldest depiction of this type of arcade and there seems to be no earlier model for this feature, not even in France.[9]

The architecture of the planters, the great profiteers of the slave economy in the Antilles, possessed a distinct style that set their houses apart from French architecture. Creole style was a phenomenon of the proprietary society that diverged from the state's interest in ensuring France and the King were properly represented in the colonies. Because of this, the royal administrator had to impose the colonial style that justified du Tertre's assertion that the architecture of government buildings was 'quite regular, with very comfortable rooms and from the outside indistinguishable from [that of] buildings in France'.[10]

One of the first visual representations of an architecture that emphasizes the austere classicist order, as was fashionable in France around 1700, was a work by Père Labat himself (built 1690–1704) (see Figure 53). Spontaneously, the façade of the Jacobin convent was reminiscent of the famous garden front of the new palace at Versailles. Labat also integrated a long gallery on the first floor with a view to the sea – it fell far short of the splendour of Versailles' galérie des glaces, but was nevertheless more than a merely functional feature. It underlined the pretensions of the clerical order not only as a beneficiary, but also as one of the wealthiest proprietors on the island.

As ornamental elements it featured an impressive cornice, holding a balustrade garnished with vases and globes, and a dramatic staircase that led to the large portal, which was flanked by two columns and crowned by an elaborate coat of arms. Not without pride, Labat arranged the building such that the whole of the ground floor was suitable for use as a residence for the occasional visit of a vice admiral or a lieutenant general: 'Thus one could consider the ground floor as a secular mansion and the upper floor as a convent.'[11]

Fort Saint-Louis on the southern coast of Saint-Domingue was meant to look particularly impressive. An anonymous draughtsman produced a series of sketches of the fortress, situated on an island off the coast of the mainland.[12] Most spectacular is the drawing for a main gate with a sun in the pediment and royal arms in the panel below, and garnished with flags, mortars, and cannons (see Figure 54); a second draft for the same gate shows an empty panel and pediment, but two obelisks at both sides. The typical guard houses (guérites), which do not appear very often in images of French fortresses in the Caribbean, display the Fleur-de-Lys on top (see Figure 55). A chapel was supposed to stand

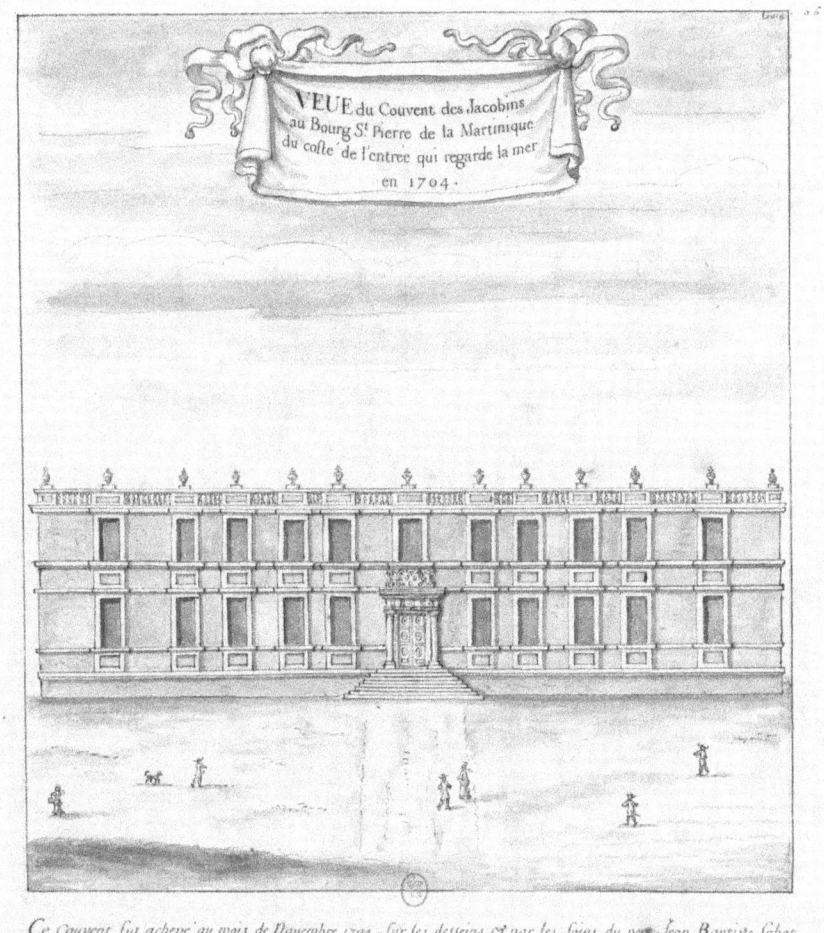

53 View of the Jacobin convent in Saint-Pierre, 1704, in Labat, *Nouveau Voyage*, vol. 4, p. 206.

inside the courtyard of the fortress. The illustrator drew a relatively modest façade with an original curved pediment with rather peculiar baroque-style spirals at the side.

The engineer in Saint-Domingue, Pierre de La Broue, responsible for the execution of the works at Saint-Louis, devoted much attention to the ornamental details.[13] Thousand of stones were sent to the construction site to build the fortification cordon and the sentry boxes. At least

[143]

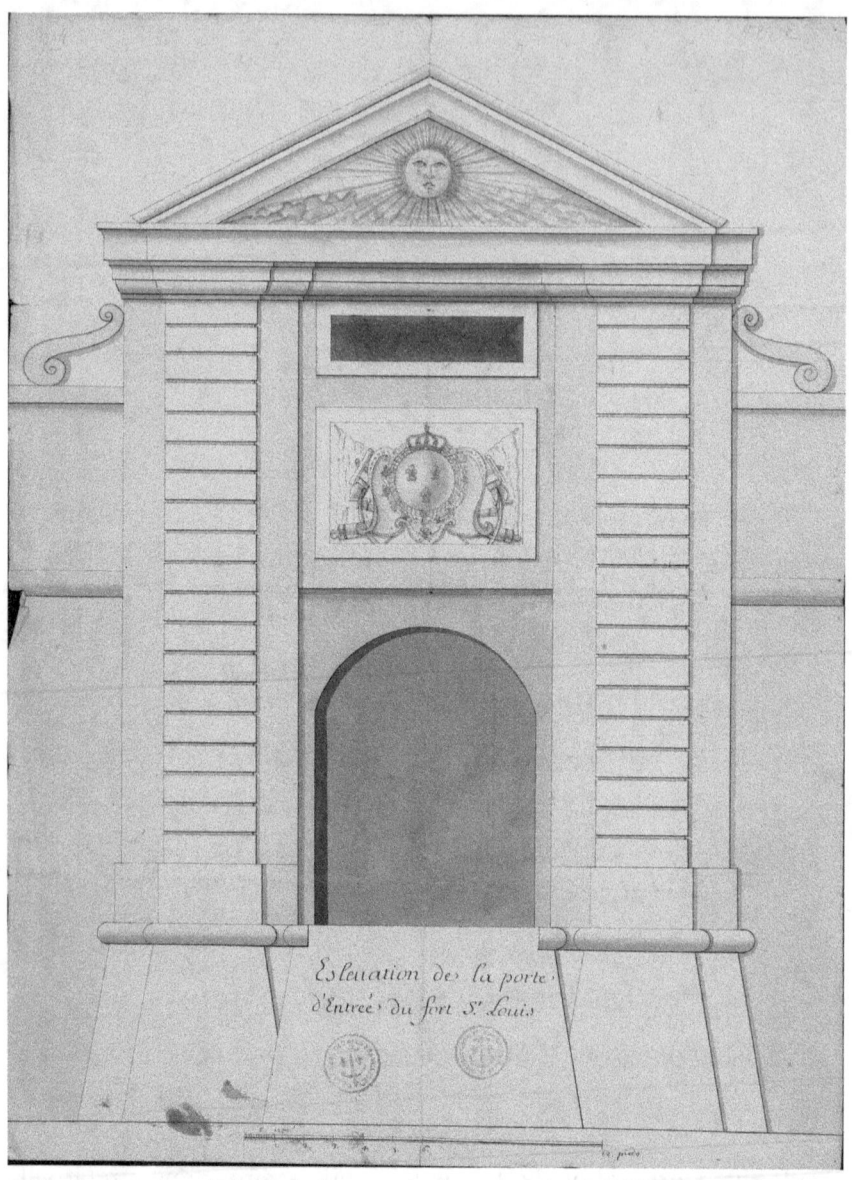

54 Anonymous [Pierre de la Broue]: Elévation de la porte d'entrée du fort
Saint-Louis, 1704 (CAOM 15DFC/824C).

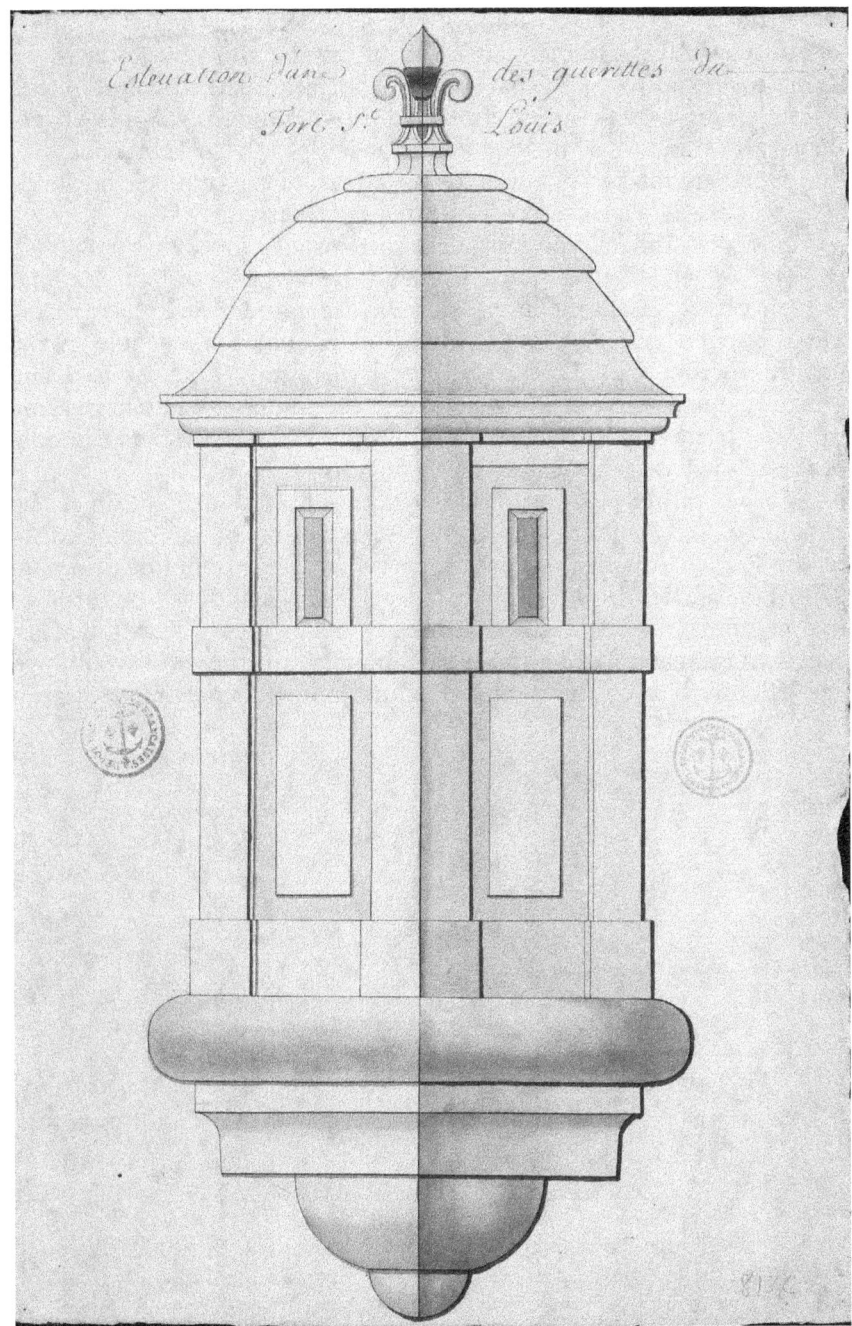

55 Anonymous [Pierre de la Broue]: Elévation d'un des guérittes de Fort St. Louis, 1704 (CAOM 15DFC/817C).

100 Africans were put to work, which the secretary of the navy, Jérôme de Pontchartrain, authorized himself in the margins of a letter by La Broue to Versailles. The space above the gate, marked black on the image, was to contain an inscription glorifying the King. For the ornaments the engineer demanded a special sculptor from France, as well as 6 experts in mortar production. His order for large cannons, however, 70 pieces in all, Pontchartrain fulfilled only hesitantly.[14]

Today only the foundations of the fortress, destroyed by the English in 1748, are visible on the little island in the bay of Saint-Louis-de-Sud; so little can be said of the impression the decorations might have made. But they were probably not meant for the eye of the people in the nearby town of Saint-Louis. The portal was remote and the gate and other ornaments were barely visible from the shore – as one can imagine from the engraving by Nicolas Ponce, where the fortress in the middle of the sea is barely noticable (see Figure 56).

Another rather elaborate case of representative decoration is the armoury of Cap Français. This town was perhaps the richest municipality in the French Antilles. Louis-Joseph de La Lance, chief engineer on Saint-Domingue, drew a design that attached significant importance to ornaments: the pediment was embellished by the royal coat of arms, spears, flags, cannons, mortars, and anchors; an iron grating adorned the balcony below, the detail of which shows it decorated with a

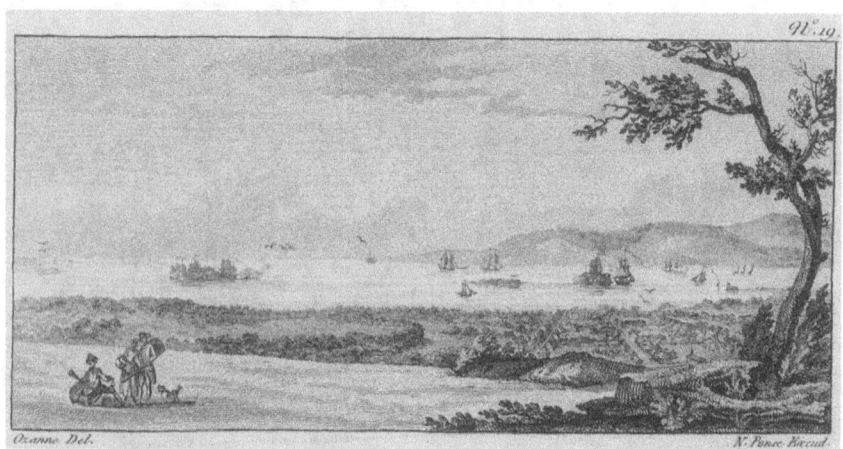

56   Nicolas Ponce, Baye et Fort St. Louis, in Moreau de Saint-Méry, *Recueil de vues des lieux principaux de la colonie françoise de Saint-Domingue*, No. 19. This work was meant to accompany Moreau de Saint-Méry's *Loix et constitutions des colonies françoises de l'Amérique*, 1784–1790.

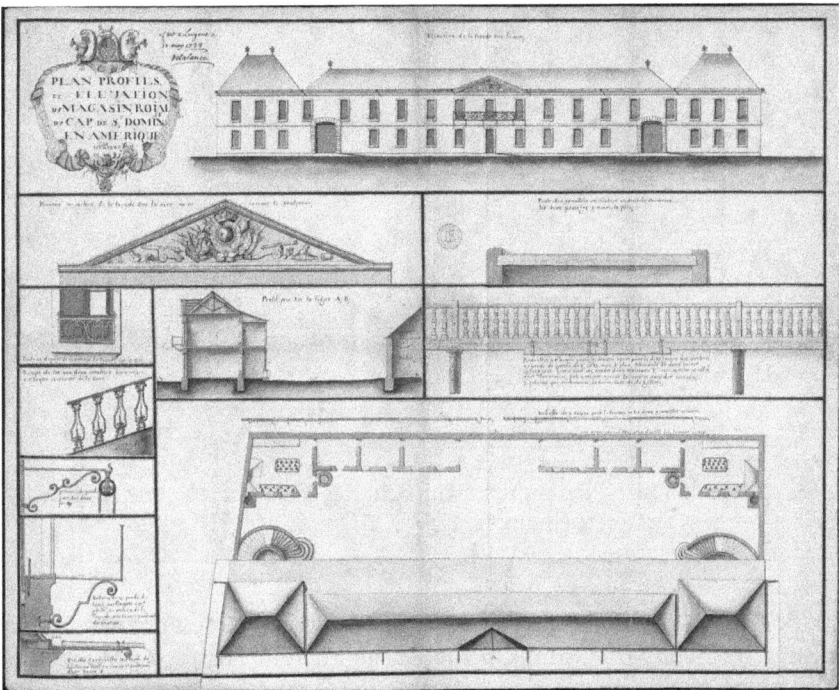

57 Louis-Joseph de La Lance: Plans, profils et élévation du magasin royal
du Cap de St. Domingue en Amérique, 1738 (CAOM 15DFC/335A).

silver-coloured finish (see Figure 57). Each first-floor window of the
seaward façade was fitted with a gilded iron grating forming the letters
'LXV' for Louis XV with the Fleur-de-Lys at the corners.

The positioning of this building was done with greater reference
to the town. It faced the sea, with its front geometrically aligned with
the city's grid layout, which became so typical of urban planning in
the Antilles during the eighteenth century.

The geometrical order that defined the proportions of the buildings
was also applied to the cities and the landscapes. Fortresses in the
Vauban style were designed with a symmetry intended to impose geometrical form on the natural topography of the islands. How this was
to be realized was, as discussed at the start of this chapter, a matter
of dispute. The island's volcanic landscape was rocky and mountainous
and difficult to transform according to the spatial ideal.

Nevertheless, engineers and planners tried to achieve this ideal
throughout the course of the eighteenth century. Fort Saint-Charles in

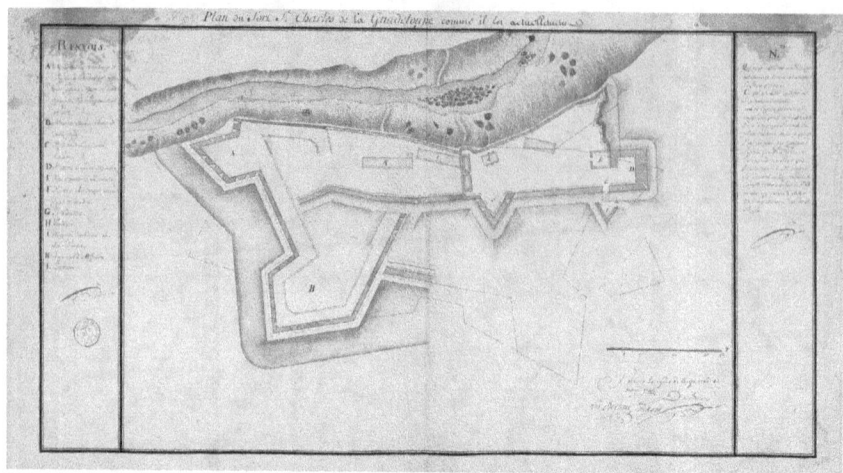

58   Louis de Bury: Plan du fort Saint-Charles de la Guadeloupe comme il
est actuellement, 1743 (CAOM 8DFC101B).

Basse-Terre on Guadeloupe was planned as a large defence work that
nearly reached perfection. On top of the ancient Fort de la Madeleine
built by Charles Houël a new fortress was established that followed
the topography, occupying only the highest ridge of the hill slope. In
1743, however, plans were continued to envelop the first fort within
a second, nearly symmetrical ring of bastions (see Figures 58 and 59).
But by 1763 nothing had materialized in this direction. Instead, only
a French-style garden had been added, geometrically planned, but still
disrupting the order of the overall conception (see Figure 60).

In 1764 a new project was conceived that put forward several designs
for a second ring of fortification. One could flip over paper flaps showing
the different technical solutions, but they all adhered to the ideal of
geometrical symmetry (see Figure 61). A plan of 1785 shows the fortress
together with the town of Basse-Terre, contrasting the constructed
space of the defence works with the cultivated spaces of the enlarged
town (see Figure 62). Military and civil construction practice thus
indicated two spatial forms that seemed to stand in opposition to each
other. A closer look at the geometry of Fort Saint-Charles, however,
reveals a discrepancy between the ideal form and the natural condition
of the rough terrain at the banks of the Galion River. Far from being
symmetrical, the fortress followed the decline towards the sea, resulting
in a profile that resembled a skewed geometrical form.

A photograph from around 1900 shows the fortress (then named Fort
Richepanse) together with the Pont Galion free of the vegetation that

[148]

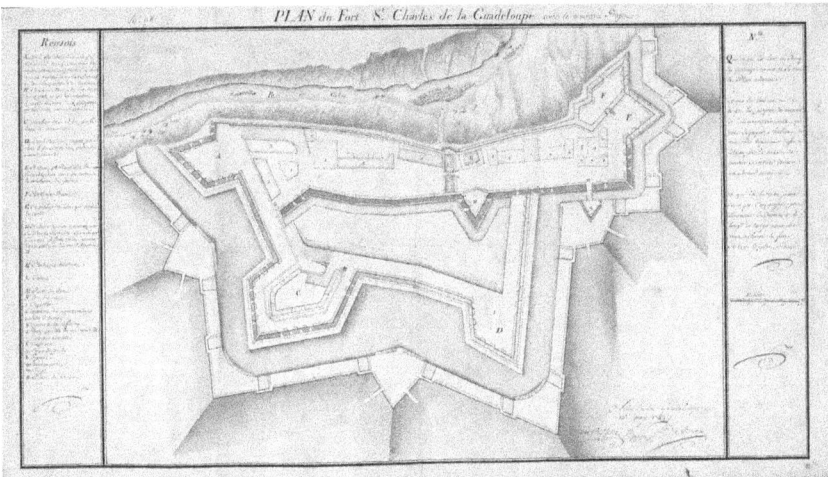

59   Louis de Bury: Plan du fort Saint-Charles de la Guadeloupe avec le
nouveau projet, 1743 (CAOM 8DFC102B).

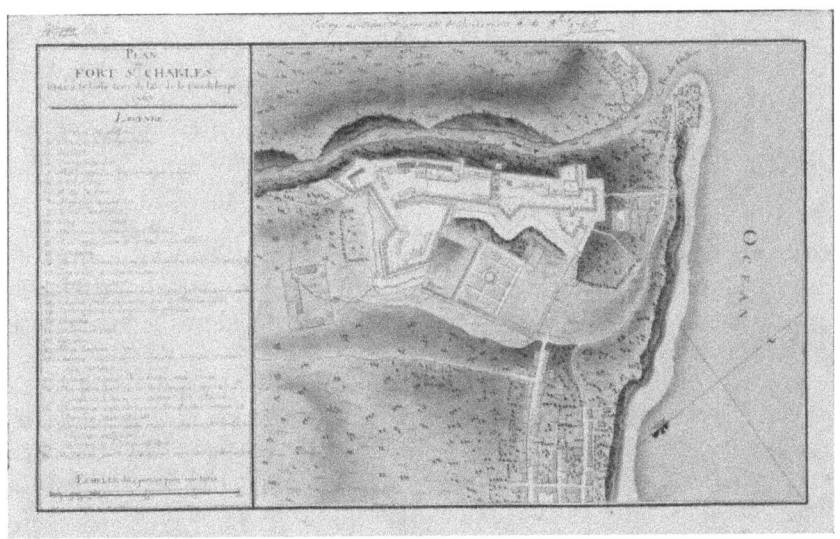

60   Vialis de Saint-Hilier de La Grange: Plan du fort Saint-Charles situé à la
Basse-Terre de l'isle de la Guadeloupe, 1763 (CAOM 8DFC128C).

surrounds the premises today (see Figure 63). Here we get an impression
of an imperfect and asymmetrical structure that does not imitate geometrical Euclidian bodies, but adjusts more to the uneven topography.
The imposition of the geometrical ideal of absolute conceptions of

[149]

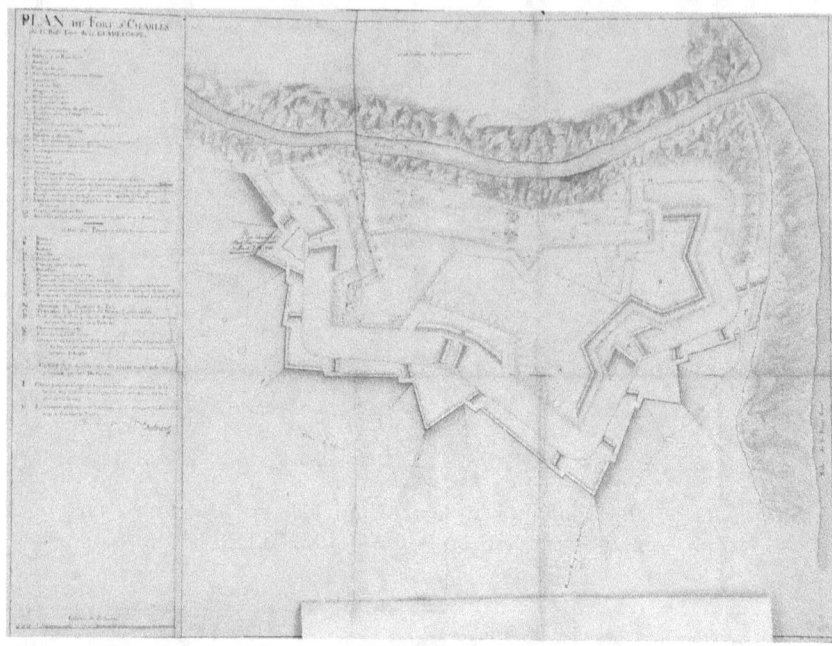

61 Henri Philippe Joseph de Rochemore: Plan du fort Saint-Charles de la Basse-Terre de la Guadeloupe, 1764 (CAOM 8DFC161A). The paper lash in the lower half of the plan could be turned to reveal two solutions for the development of the outer cordon of the fortress.

space on urban environments was, however, achieved in the cities of Saint-Domingue. Cap Français and Port-au-Prince are the most prominent examples of the city grid structure that was devised in the eighteenth century (see Figure 64).

A further example is the town of Fort Royal on Martinique, planned as early as the seventeenth century. But, as so often with large projects, the realization of the concept was difficult. What may have looked very attractive on coloured maps in fact remained a deserted area for many decades. In 1698, the plan by the engineer Caylus for a new town of Fort Royal, where only a Capuchin convent and a small church stood in the middle of some marshland, provided for the supply of water via a stone aqueduct, canals connecting the harbour of the Carenage with the Cornet River, a Jesuit college, a hospital, and store houses (see Figure 65).

In 1764, Fort Royal remained a very small town, a plan showing only scattered houses, a small administrative building for the intendant, and a few temporary wooden bridges over the canal (see Figure 66).

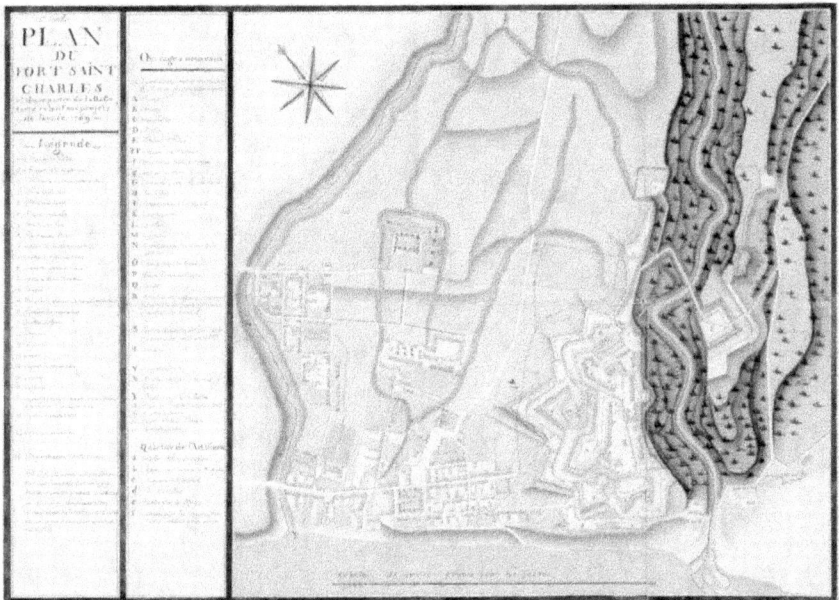

62 Paul Edme Crublier de Saint-Cyran: Plan du fort Saint-Charles et d'une partie de la ville Basse-Terre, relativement aux projets de 1785 et 1786, 1785 (CAOM 8DFC400A).

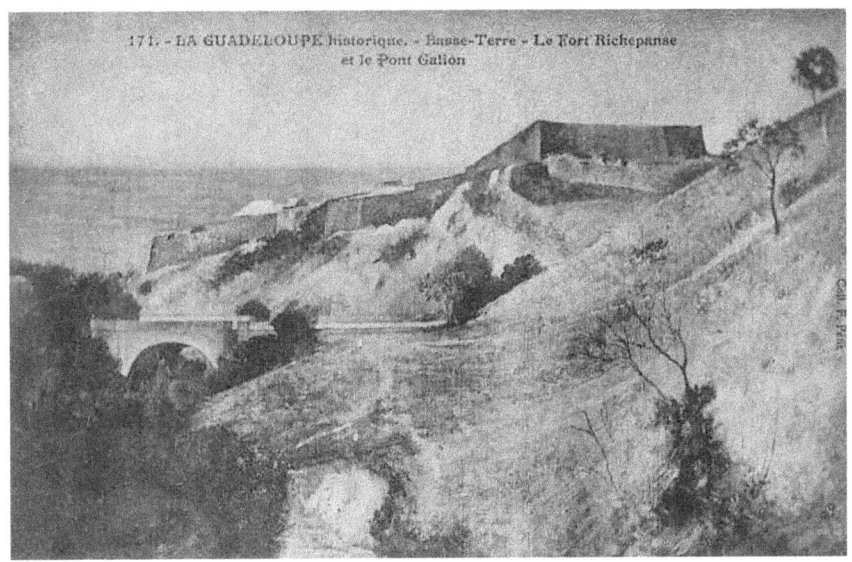

63 Postcard of Fort Saint-Charles, then Fort Richepanse. In the foreground is the Pont Galion built in the 1770s.

[151]

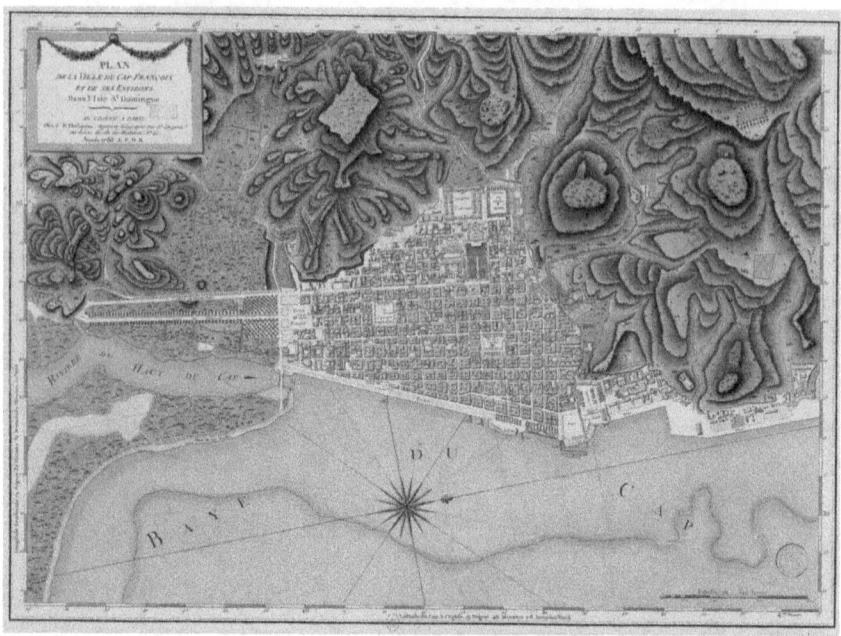

64 Plan de la Ville du Cap François et de ses environs dans l'Isle de St. Domingue, Paris: Phelipeau, 1786 (BNF Paris, GE SH 18 PF 149 DIV 4 P22 D).

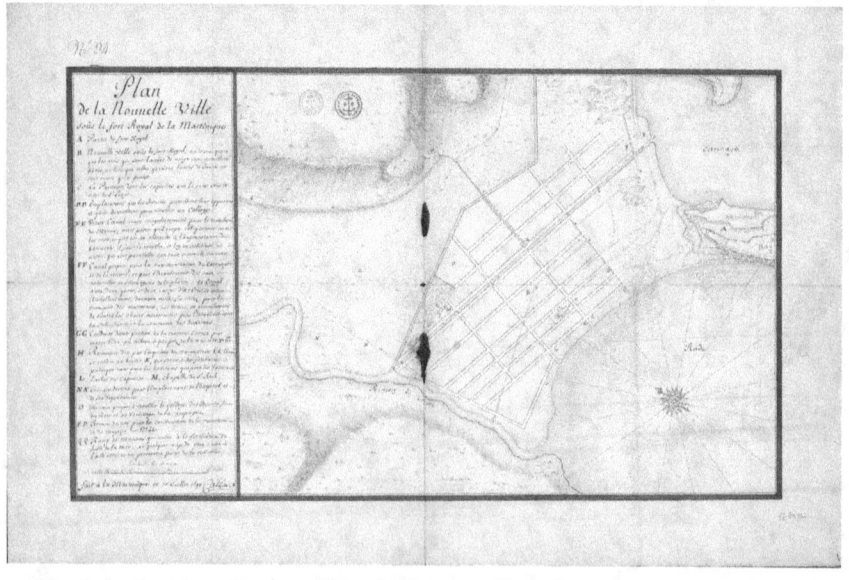

65 Jean-Baptiste Caylus: Plan de la nouvelle ville sous Fort Royal de la Martinique, 1698 (CAOM 13DFC86bisB).

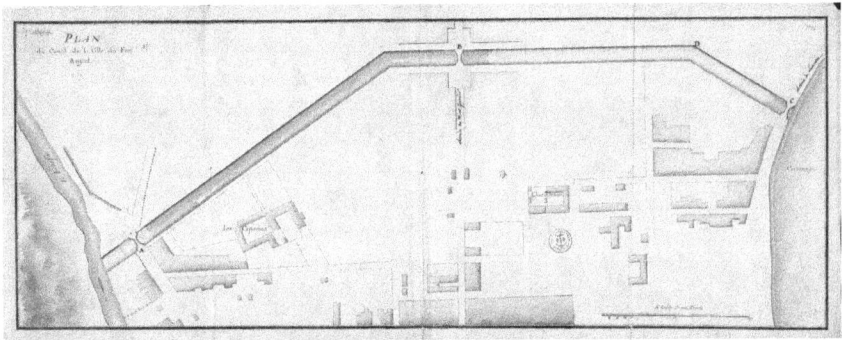

66 Henri de Rochemore: Plan du canal de la ville du Fort Royal, 1764 (CAOM 13DFC225A).

Plans of colonial cities of this period are difficult to interpret. On the one hand, they seem to convey reliable information in presenting a detailed legend identifying individual buildings. On the other hand, there is a lack of detail for most of the residential areas, usually uniformly displayed as coloured blocks that imply a dense urban population. A series of engravings by François Denis Née after sketches by Chevalier d'Eperny from ca. 1780 shows the town of Fort Royal from three perspectives with only few buildings and facilities (see Figures 67, 68, and 69). Their loose order contradicts the impression of the geometrical maps as well as the supposed size of the town. The images allow another observation that the boundary between urban and rural space is very indistinct. The town seems to gradually merge with the agricultural areas of the plantation economy. Indeed, a drawing of the area of Fort Royal from 1702 by the naval officer and botanist Roland Barrin de La Galissonière shows how the scattered houses escaped the urban confinement of the canal and connected with the surrounding fields (see Figure 70).

A similar case can be made for the relationship of rural and urban space in Port-au-Prince and indeed has already been made.[15]

But juxtaposing rural life in a space cultivated by Creole elites with city life in a space constructed by colonial elites belies the interconnections that existed between the different groups participating in the construction of urban spaces. Colonial cities such as Port-au-Prince and Cap Français opened up opportunities for participation in culture, knowledge, and, of course, politics. As James McClellan has argued, the urban context of Cap Français allowed the emergence of cultural institutions such as the theatre and the printing press, which introduced knowledge about literature and science, like the periodical *Affiches Américaines*, thus leading eventually to critical potential in the Creole-colonial society of Saint-Domingue.[16]

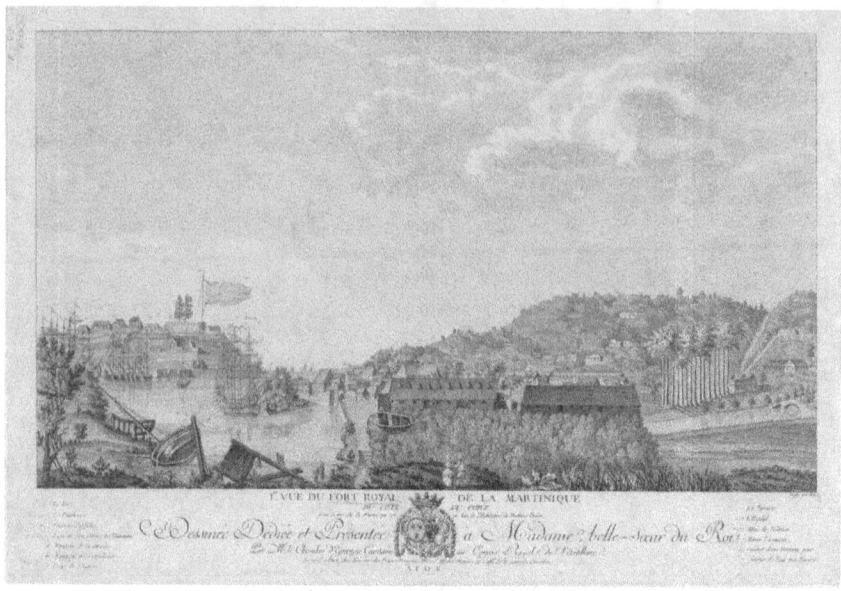

67 Chevalier d'Eperny/François Denis Née: 1er vue du Fort Royal, ca. 1780.

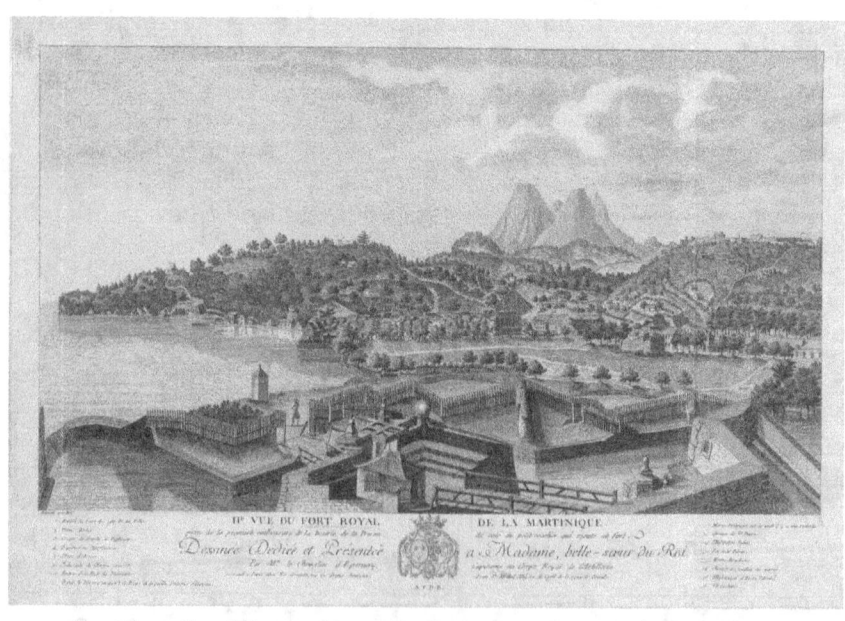

68 Chevalier d'Eperny/François Denis Née: 2e vue du Fort Royal, ca. 1780.

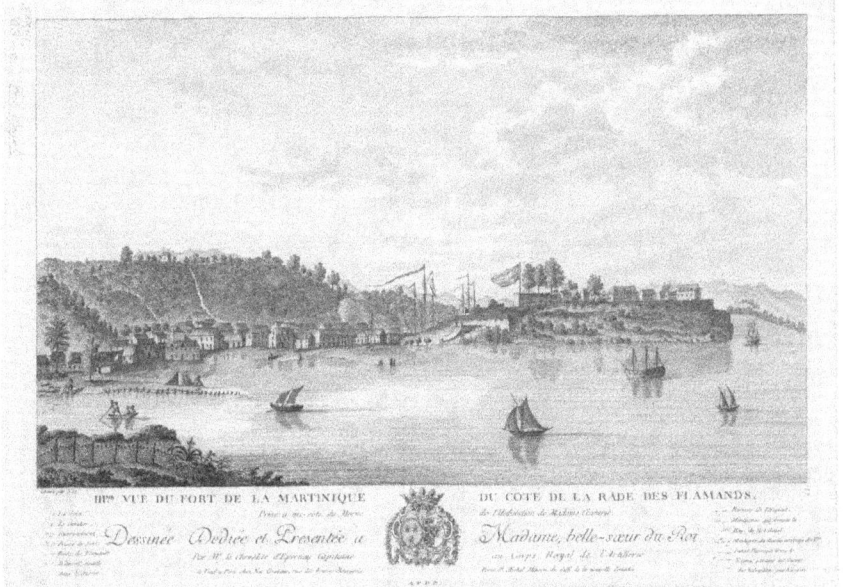

69   Chevalier d'Eperny/François Denis Née: 3e vue du Fort de la
Martinique du coté de la rade des flamands, ca. 1780.

Trevor Burnard and John Garrigus recently described the colonial
cities in the Caribbean, borrowing a term from Fernand Braudel, as
'electric transformers' or as urban spaces that 'were also liberating
spaces for free and enslaved people of colour, who often found more
autonomy, a wider set of social relationships, and greater earning power
than in rural areas'.[17]

There were, indeed, positive as well as negative side effects of the
effort to impose a French style in the planning and execution of colonial
cities, fortresses, and large buildings, creating spaces with economic and
military functions. The classic modernization narrative has always seen
something good coming from the construction of hospitals, libraries,
schools, and universities, without delving too deeply into the complexity
of contradictory conditions under which modernity arose. But there
is another point to be made here. In all of the building efforts, from
slave hut and plantation mansion to the large dominating fortress
and the town complex, there were many people involved – not just
those who initiated the project, and planned the financial and material
resources, but also those who knew how to build a wall or to prepare
the mortar, along with those whose only contribution was their physical
strength.

[155]

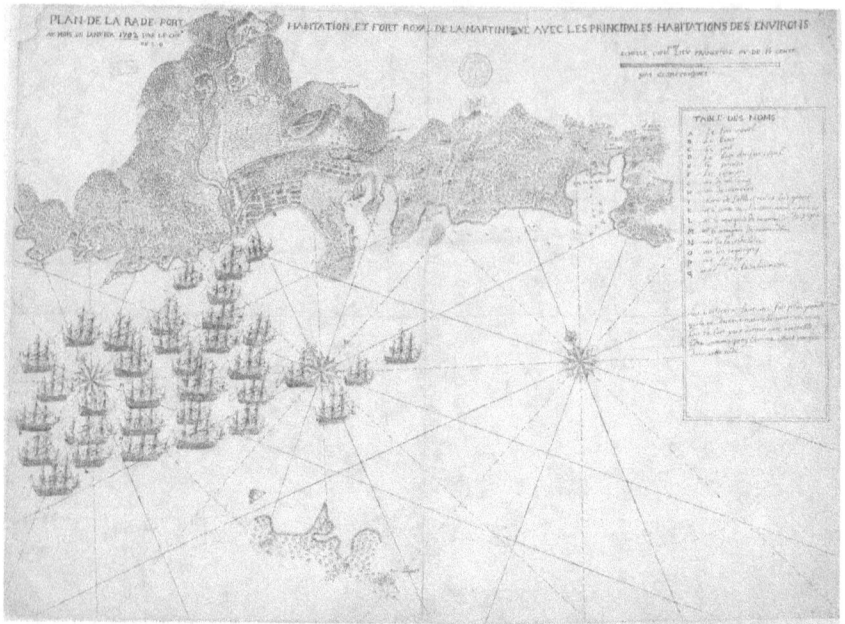

70   Roland Barrin de La Galissonière: Plan de la rade, port, habitation
et Fort Royal de la Martinique avec les principales habitations des
environ, 1702 (BNF Paris, Département des Cartes et plans,
GE SH 18 PF 156 DIV 5 P 6 D).

It is they who must be seen as the historical agents for creating all
these structures, with their potential to hold so many different meanings.
As a combined social cohort they were, of course, marked by their
differences of rank or class, being slaves or masters, public administrators
or private contractors, skilled or unskilled workers, overseers or labourers.
But in the work process they formed a unit that was interrelated through
the task at hand. The materiality of the constructions bound these
very different people together and combined them as a community.
And because these buildings created affective spaces, public or private,
they could reflect emotionally back on those who were involved in
building them and, in time, those living with them for many years and
sometimes decades or centuries to come.

Production of emotional spaces

A last example serves to illustrate this process of physical construction
and the formation of an emotional community in the Antilles. The
following memorandum is taken from the administrative documents

of the National Archives of France and concerns the plan to build an aqueduct from the Turgeau and Chavanne rivers to a fountain at the Place d'armes in the centre of Port-au-Prince.[18] The author, a Monsieur Hesse, the royal engineer responsible for Port-au-Prince, gives a very detailed account of how this project was to be executed. It details the people who were needed, the proper choice of materials, how to prepare them, the different experts necessary for special tasks, the necessary hierarchical and logistical organization, and the costs and the source of finance for the whole thing.[19]

It is not entirely clear whether the project was executed as planned. The fountain itself was dismantled after a few years. But the project is a good example for two reasons: first, it shows how the practice for large building projects that was introduced in the Antilles in the seventeenth century became something of a routine that was applied not only to military, but also to civil public projects. Second, though the canal project is only one of many, it is well documented, unlike the majority of large building projects in the Antilles. The memorandum is technical in its detailed description of the undertaking, but vividly portrays the everyday situation of a construction site in the colonies.

Hesse begins by describing the material that he proposes to use for the project: dressed stone (*pierre de taille*) for the canal, mortar of sand and lime, and cement for the inside of the trough. The fountain in the centre of the Place d'armes was to be built of brick. It was to have a large basin, four troughs at the sides, and, supported by cornices and ledges, a pyramid on top (see Figure 72). The heavy work, the removal of earth, stones, tree roots, etc. was to be organized by a contractor and a royal engineer. The contractor had to guarantee that at least twenty-five African or Mulatto workers would be employed every day; the engineer had to supervise fifty dayworkers (*mainœuvres*) at different locations. A deadline for the completion of the project was scheduled for after two years, provided there were no accidents caused by extraordinary circumstances like, for example, an earthquake. If the deadline was missed due to low-quality materials or assemblage the contractors were to be held accountable.

The chief engineer Hesse obviously felt obliged to instruct the project leaders with even more detail on the materials. Stone for the basin had to be imported from France, from Crazannes to be precise, a well-known quarry for a white limestone in the vicinity of the shipyards of Rochefort. For the masonry of the canal, rock and stone were to be taken from the local vicinity, while bricks had to be of the best quality imported from Provence. Roches à Ravet, a porous and ferriferous rock that could be found on the island, and river pebbles could be used for lime burning. Everything had to be prepared very carefully and thoroughly: mortar

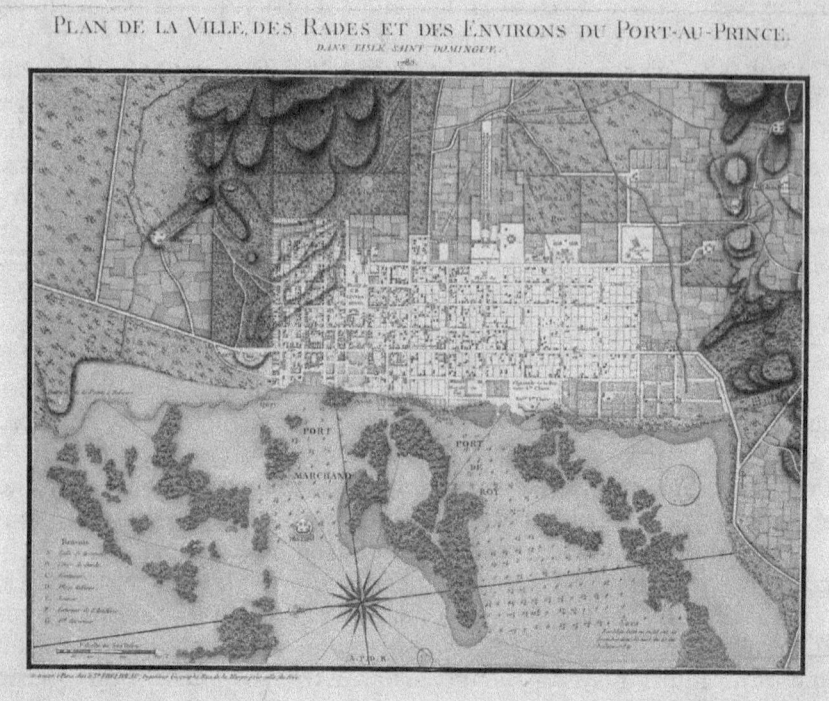

71 Plan de la Ville des Rades et des Environs du Port-au-Prince (Paris: Phelipeau, 1785) (BNF Paris, GE SH 18 PF 149 DIV 4 P 14 D).

had to be made of one-part lime and two-parts sand from the riverbeds in the Ravine aux Chats or the Chavanne, two nearby rivers in Port-au-Prince. Clay was to be added, firmly stirred, and with as little water as possible. The contractor and engineer were forbidden under any circumstances to take sand from the sea. Wood for carpentry should be of Acajou, a kind of mahogany also of excellent quality and showing no knots; iron was to be checked for possible fissures; hinges, locks, and bands had to be used for the doors of the maintenance hatches. These were to be coated with oil paint.

Hesse expected everything to be executed in the best fashion and quality. The price was fixed at 90,298 colonial livres, which was to be paid on completion of the work. The house and plantation owners who profited most by the new water supply paid the largest part of the costs; the King contributed the rest.

The canal was part of a larger hydraulic system for the city of Port-au-Prince, which had no access to fresh water of its own, but eventually ended up having one of the best water systems in the Antilles, second

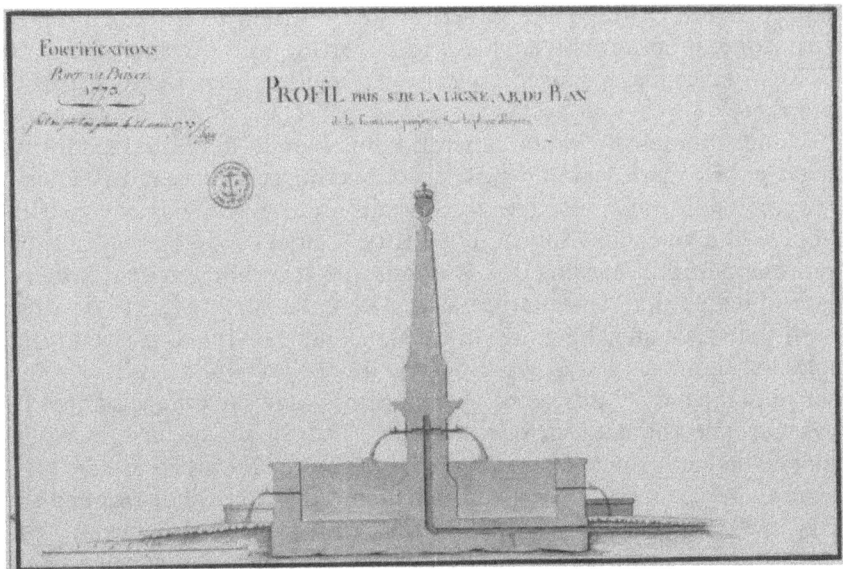

72 Hesse: Profil pris sur la ligne ABC du plan de la fontaine projetée sur la
     Place d'armes, Port-au-Prince, 1773 (CAOM 15DFC621C).

only to that of Cap Français.[20] Three canals accessed three rivers, the
Chavanne, the Thurgeau, and the Charbonnière, which converged in
a surge chamber (*Château d'eau*) and a central basin of distribution.
Three water conduits went from the central basin to the barracks, the
royal hospital, and to the Place du Gouvernement, from there diverging
to the garden of the house of M. Hesse and, along the Rue de Rouillé,
to the Place d'armes and the large basin serving as a public washing
area. Another conduit connected the surge chamber with the garden
of the intendance, another basin and a fountain were situated on the
Place de l'Intendance, and via the Rue des Fronts Forts and the Rue de
Bonne Foy water was directed to the Place Valière and the Aiguade at
the merchant port (see Figure 71).

   According to the chronicler of Saint-Domingue, Saint-Méry, the
planning for this hydraulic system began as early as the 1730s. A first
canal was built in 1743, but did not satisfy the demand for water in
industrial measures necessary for the production of indigo, chalk, and
the farming of manioc. Also, the water it carried was not clean enough.
The governor-general commissioned several engineers and land surveyors
in 1761 and in 1764 to find a solution for a better and larger water
supply. Surveyors and land proprietors exchanged reproaches against
one another: the surveyors were accused of insinuating possible land

[159]

expropriation through the government in letters to the planters. In turn, some proprietors were fined for diverting water to their fields. In 1770 an earthquake caused crevasses to open up that led to a further shortage.[21]

Under the new governor-general, Louis-Florent de Vallière, another attempt was made to build a canal connecting the Turgeau River with the city. Saint-Méry describes the curious scene at the start of construction, when one morning in 1773 fifty soldiers together with their commander, the Marquis de Laval, and the governor-general, Vallière, assembled at the construction site. The governor made the first cut with the spade and the work on the canal started. It was a fête for the soldiers, Saint-Méry says, working over a period of four months without pause and salary. Vallière often stopped by and encouraged them by his presence and paid for refreshments. The canal was built of wood, snakewood to be precise, which grew at the site, as only local materials were to be used. When it was finished and the first water reached the city, Vallière solemnly drank the first glass, everybody cheered, and the governor rewarded the soldiers with the promise of another canal for the regimental barracks.[22]

In July of the same year, the water supply was still not sufficient and the construction of the canal planned by the engineer Hesse commenced. It was finished in December, connecting the government, government square, and barracks to the water system. In October 1773 construction of the canal began to tap the water sources of the Chavanne plantation, which was completed in 1776. The fountain Hesse planned for the Place d'armes was not the only one. He also estimated the costs for one to be built in the square in front of the intendance (70,000 colonial livres), one for the Place Vallière (40,000 colonial livres), a watering point (aiguade) at the Quai du Rohan (25,000 colonial livres), and another one at the Esplanade of the Sainte-Claire battery (66,000 colonial livres), in total 205,000 livres. 60,000 livres were to be paid by the propriators; the merchant captains were to be asked for 25,000 for the aiguade at the pier (see Figure 73), and the rest was to be discharged by a tax collector (receveur).[23]

However, none of these fountains were realized initially. In 1776, the American War of Independence interrupted all public works that were not of military relevance, but after 1787 work on the fountains recommenced again. Costs, however, rose and quadrupled to more than 800,000 livres – a sum, Saint-Méry added, that would have been much higher were it not for the slave workforce in the King's possession (l'atelier des nègres du roi). Three sources provided the several fountains and watering points at Port-au-Prince with water by dint of three stone-built canals. The system was maintained at great expense: an ordinance

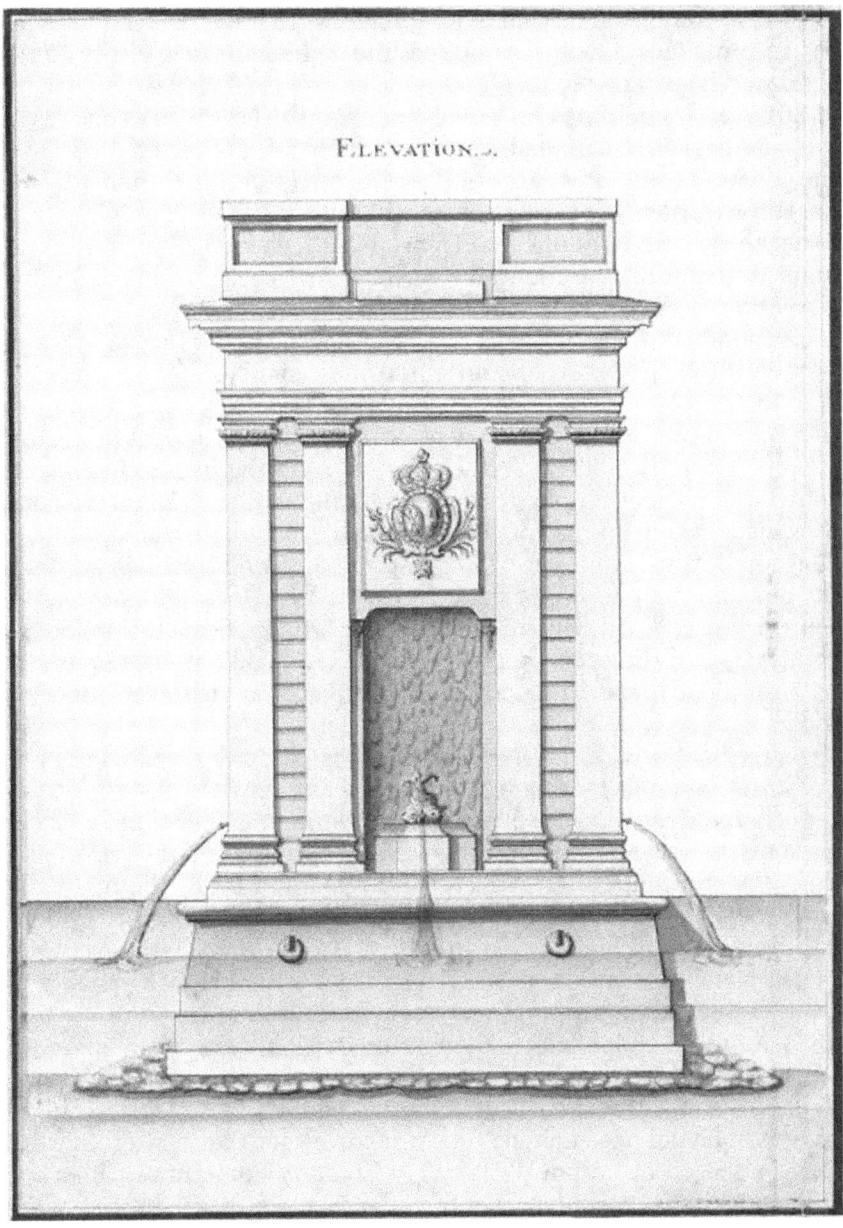

73 Jean-André du Coudreau: Plan, profil et élévation d'une fontaine à faire sur le bord de la mer du Cap dans une des cales du quai pour l'aiguade des vaisseaux, 1747, Detail (CAOM 15DFC354C).

of 1788 appointed an official hydraulician, with a salary of 4,000 livres per annum. This officer was placed under the authority of the Corps royal du Genie, created in 1776.[24]

But even though the city now had one of the best water supplies of the islands – an accomplishment for which Saint-Méry credited only the governors and intendants of the respective administration – it was not universally used. Most people still preferred to use rainwater, and under the government of César Henri de La Luzèrne no one in the city drank water from the fountains. Instead, Africans continued selling water on the streets – a practice that had not been mentioned by Saint-Méry before, but which seems to have been the major means of water supply in the city before the arrival of the canal system.[25]

Saint-Méry's account leaves us with a favourable impression of the French administration successfully installing a working water supply system in Port-au-Prince. Not only did the hydraulic projects contribute to creating a better and more functional city infrastructure, they also created a sense of community. The example of the soldiers, allegedly joyous about the work despite bad working conditions and the fact that they were not paid, reveals that the account was biased. The omission of Africans in crediting the workers responsible for the completion of the canals and fountains also highlights the prejudice that the French somehow could plan and accomplish everything on their own.

As we have seen from the analysis of the agencies involved in large construction projects, this is not true. In fact, in the daily routine of the construction work, the planning of the structures and of the finances was only as crucial as the manpower, the expertise, and the knowledge that was contributed by African slaves. In the case of the hydraulic works of Port-au-Prince, many of these workers were expert masons and limeburners, and skilled in many of the other tasks that were indispensable for building as complicated a structure as a canal. The final achievement was shared by everyone involved. It was, by this logic, a building planned by the administration, paid for by planters, but built by slaves and soldiers.

Interestingly, the emotional bond the work on these projects might have formed did not embrace the Africans or, for that matter, most of the population of Port-au-Prince. The soldiers, sometimes called the 'white slaves' (nègres blancs) of the colony, might have cheered at the prospect of having their own water supply.[26] But most of the inhabitants, the Africans, the majority of the island's population, rejected the innovation. They stuck to their own traditional practice of water supply.

The white stone that was used for the fountain at the Place d'armes seems to have been intended as a material reference to France maybe even to La Rochelle, the 'white city' (*la ville blanche*), which was built of the same stone. It represents the colonial ideal of imposing the metropolitan style on the city's population. This is expressed, too, in another passage in Saint-Méry's account on the aesthetic dimension of the cities of Saint-Domingue. Writing about the luxury on display in Cap Français he says that it is pursued like a cult, giving out its pleasures and vices, 'like in a centre'.[27] But this expression was not an allusion to a colonial metropolis, that is, an imperial capital like Paris, in the sense of Captain Alexandre Berthier's bon mot that Cap Français was 'the Paris of our colonies'.[28] It rather alludes to one of the peripheral centres in France, like Nantes, Bordeaux, or Le Havre, that bear more resemblance to the descriptions of Cap Français or Port-au-Prince.[29]

Colonial cities, however, never really adapted to these aesthetic improvements directed from above. The general picture that contemporaries perceived was rather that of a polychrome, brightly coloured, and heterogeneously styled architecture. There are several writers whose impression of the cities of the Antilles was not favourable. In Michel Etienne Descourtilz's opinion, for example, only those settlers who had never left Saint-Domingue were able to compare Cap Français to Paris, since 'those two cities have nothing in common'.[30] And Port-au-Prince appeared to the artillery officer Desdorides as badly constructed, its houses scattered, and the streets crooked.[31] The whole city gave the impression of an army camp in disarray – 'un camp de tartars' – and several accounts underlined the deviation from the ideal city plan.[32]

A city like Port-au-Prince had, therefore, more than one style representing different spaces. They did not exist separately from each other, but more in relation to each other. Colonial and Creole styles coexisted sometimes distinctively and sometimes indistinctively. A view of Port-au-Prince, drawn before 1785 (see Figure 74), shows the city as it looked from the other side the bay, the houses seemingly of stone (which they were not, as most of them were build of wood), not aligned to a rectangular grid pattern, but collected around the two administrative buildings of the governor and the intendant.

The anonymous artist highlighted the governor's building with its two wings, portico, and double staircase. A fence with an elaborate gate enclosed the square in front of the building. Less extravagant was the seat of the intendant, being only a slightly larger edifice than the surrounding houses and having only a lightly ornamented gate set into the fence in front of a smaller square. The church next to it was rather small, too. The only other signs of colonial rule were the gigantic white flags flying above the three fortresses guarding the harbour of Port-au-Prince.

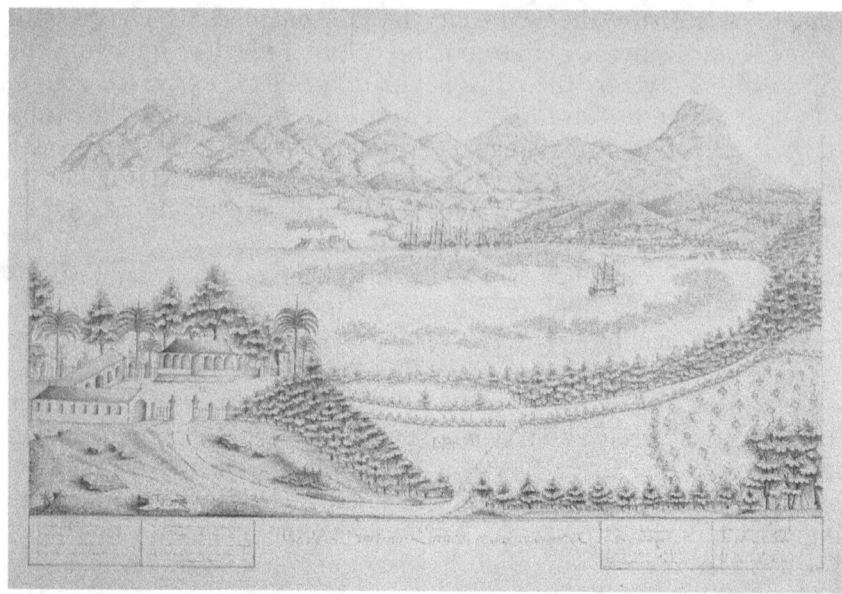

74 Vue de Port-au-Prince et ces environs, ca. 1800, dated according to the Boston Public Library catalogue, but probably before 1785 since there are no buildings to the right of the government building that were added on a map of Port-au-Prince of that year.

The Creole style of the plantations dominates the picture. The plantation of M. Martissance, a sous-commissaire de la Marine, is placed prominently in the foreground. Its architecture displays the typical gallery around the main house, the modest ornamentation of the entrance gate, and a garden stocked with different local trees. Among other plantations and crop fields surrounding the city one can discern the African quarters of Arracs and Varreux outside the town. The general picture of Port-au-Prince as it is sketched here combines different urban spaces in close vicinity to each other. The French colonial architecture, however, does not overshadow the Creole style. Only the African style, the villages, and the slave huts that occupied the ground behind the government building remained hidden on this picture.

Urban spaces were contested and could lead to outright conflict. The Haitian Revolution that followed shortly after the completion of the water supply system brought up not only the social question of the slave economy, but also questions of cultural and material identity. The new rulers of the island continued to use materiality in the form of large building projects that could create political spaces and emotional

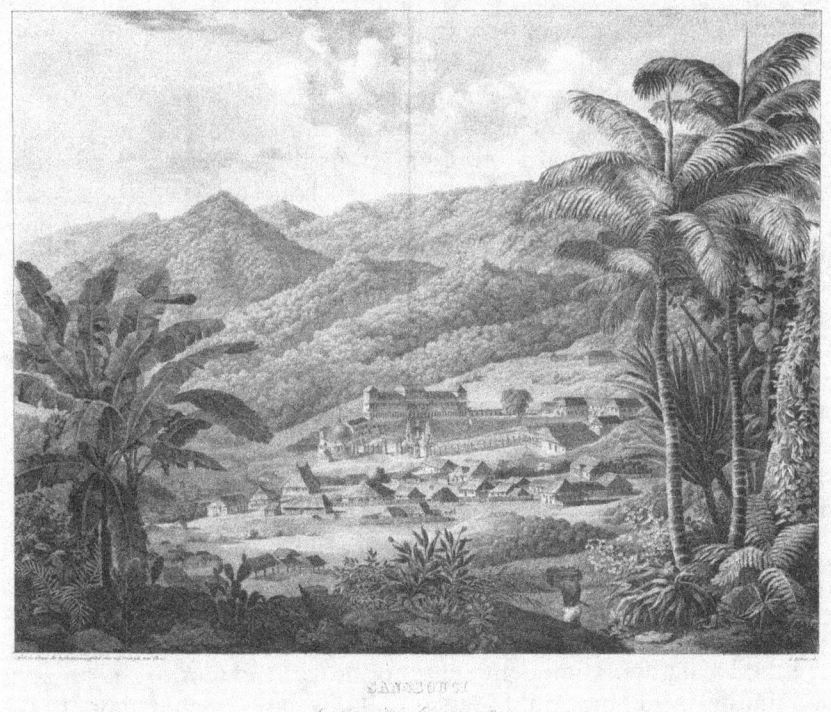

75 Sanssouci: Residence of King Christoph of Haiti, in Ritter, *Naturhistorische Reise nach der westindischen Insel Hayti.*

communities. The Haitian government under Henri Christophe authorized an enormous programme that was to demonstrate the capabilities of the new Black Republic to the world.

The Palace of Sans-Souci (built 1810–13) and the royal church near the town of Milot in the northern part of Haiti were monuments, as Pompée Valentin, the Baron of Vastey, wrote proudly in 1819, 'erected by descendants of Africans, [which] show that we have not lost the architectural taste and genius of our ancestors, who covered Ethiopia, Egypt, Carthage, and old Spain with their superb monuments' (see Figure 75).[33] The Citadelle LaFerrière (built 1805–20), very nearby, represented the Haitian ambition to fortify their territory, the same way as the French began their colonial project in the Antilles two centuries earlier. Together with Fort Jacques and Fort Alexandre it was part of a defensive system achieved with the help of the same men and women that were so experienced in building fortresses: Africans that had finally gained freedom from their masters, but used the same means of asserting their collective identity.

French ambition to create a homogenous imperial style on the Antilles thus led to both intended and unintended consequences. Architectural elements, ornaments, and representations did perhaps contribute to the establishment of a certain emotional bond between the islands' population and the French government, especially to the royal body politic. But the successful slave revolt in Saint-Domingue and the foundation of the new state of Haiti demonstrated the potential of the same architectural symbolism to forge a new independent identity. Unintentionally, the French building activity in the Antilles served as an example of how to build a state and establish authority. The former African slaves and founders of Haiti used the knowledge, expertise, and experience they acquired during the previous centuries to continue the construction projects in order to defend a territory and to create a new affective space for the freed independent nation.

Was the building activity in the Antilles typical of the French empire? How did large construction projects influence the establishment of colonial authority in different locations in Canada, West Africa, India, or on the Mascarenes? Were there similar ambitions to imperial style? Did the local population contribute to the same extent to the practice of building? And how could French building activity contribute to the formation of emotional communities as well as to the formation of a new identity beyond the period of colonization? To find answers to these questions a comparison with other places where the French engaged in colonial building activity might be of help. In looking closely at the building projects in Louisbourg on Nova Scotia, in Saint-Louis, Gorée and Galam in West Africa, and Pondichéry in India more insight is gained into the commonalities and differences that form identities in a global *and* local setting.

Both settings were responsible for the formation of 'glocal' identities in the French empire that combined the imperial ambitions of the colonial administration and their Creole appropriation in times of revolution, decolonization, and post-colonial state building. The variety of materials and agencies that was characteristic for the building activities on the French Antilles is even greater on the global scale. The workforce at the Coromandel Coast in India was socially very different to that of Senegambia; the material available to engineers on Cape Breton Island distinct from that in the Antilles. But nonetheless there was an effort to standardization despite the variety of people and things involved in the building process. The scope of French ambition to imperial style, however, was global and merged into a local appropriation of that style.

This mutual dependency of globality and locality is a typical trait of the 'glocal' character of the French empire.

Local appropriations of empire

To sum up, it remains to conclude that the examination of the conditions in which large building projects were carried out in the French colonial realm discloses some striking differences, but also some common features. As in the case of the Antilles, the interplay of colonial and Creole style in Pondichéry and Senegal was responsible for the emergence of a form of empire that allowed different groups to identify with imperial architecture. In Haiti, the local appropriation of empire was nearly complete when the former slaves not only took over the territory of the former French colony, but also continued the practice of defending and enclosing the new nation by means of large construction projects. Here, the African slaves simply took what they already possessed from years of providing muscle power, expertise, and knowledge to the material manifestations of their former masters' imperial ambition. In fact, the large buildings and the spaces they constituted were the ideal target of a revolution that occupied them to change the social order of the island.

But Saint-Domingue was the only colony the French 'lost' due to a revolution. The other islands of the Antilles remained as parts of the colonial space; together with Guiana, Réunion (Ile Bourbon), and Mayotte they are still integral parts of France's territory as formal overseas departments and regions (*Départements d'outre-mer et regions d'outre-mer*). In part, the persistence of empire on Martinique or Guadeloupe is a result of the meaning 'empire' has acquired on these islands. The Creole style that the inhabitants of these colonies had developed for themselves represents more an economic order of the haves and have-nots than a colonial or an imperial order imposed by a dominant metropolitan centre. Those who participated in the construction projects of the fortifications of Fort Royal or Basse-Terre, the mansions of rich planters and governors, and the public works of bridges, hospitals, and canals were emotionally attached to the edifices that constituted the physical identity of the islands. Naturally, this emotional attachment could have been caused by the traumatic experience of slavery as well as by the elevating influence the affective language of architecture was able to exercise over its environment.

In contrast to this Creole-colonial style, the large building projects conducted in Pondichéry had to compete against a dominant local culture that exercised an even greater influence over French imperial ambitions than the Creole style did in the Antilles. The engineers and architects of French India had to devise a particularly strong display of affective royal symbols in order to impress the local elite as colonial rulers. Dupleix's governmental palace, the gates of the city walls, and the monumental Fort Louis were rather an expression of accommodation to the local landscape of power than an imposition of a distinct French idea of empire. Similar to Louisbourg in North America, Pondichéry was a fortified city, too, but Europeans were a long way from constituting the majority of the population. Here, the architecture was intended to please Europeans, Indians, and Tamils alike, and even led to the appropriation of French style by the Brahman elites of the city. Though the imperial project of French India was halted by defeat in the Seven Years' War, 'French complexions' were leaving their traces in the *métis* community of the 'white town' Pondichéry for many years to come.[34]

The African component of empire was particularly influential in the Caribbean, but it also found expression in the formation of another variant of Creole-colonial style in Senegambia. Most conspicuous is the absence of a dominant element of French style in the colonial settlements of Saint-Louis and Gorée. Instead, the presence of architectural features from several West African societies emerge predominantly in the outer appearance of government buildings, the houses of the Atlantic Creole elite, and the few public works in the urban context of the region. The aesthetics of fortification played a minor role in the proto-imperial effort to militarily secure an area for French trading interests in the Upper Senegal Valley. As the fortresses of Senegambia were chronically underfunded, understaffed, and also inadequately provided with materials, they remained only precarious posts, always in danger of being overwhelmed by natural or enemy incursions. But a physical French presence nevertheless left its mark on Senegambian society and reconfigured the Lusoafrican Creole setting.

In contrast to this Creole type of empire in Africa, India, and the Antilles, the North American context conditioned the segregated identities that characterized the colonial towns of Ile Royale and Nouvelle France. Louisbourg is but one striking example of planning a French settlement that did not rely predominantly on non-European agency in the effort to realize large building projects. In fact, Louisbourg had been conceived from the beginning as a planned city in an environment that was not too different from metropolitan France. People and material emanated from the French Atlantic coast, which enabled the colonial administration to transfer European models of city planning over the distance of the ocean. Louisbourg became emotionally much more attached to France, its social habits, and its political culture. Practices of local appropriations of empire were noticeable, too. But the administrative elite exercised them to a large degree only in order to emancipate the interests of the local colonial establishment against the intrusions of a not-so-powerful colonial machine of Versailles. Materially and socially, however, the whole of New France appeared to be more French than other colonies. Few indigenous building practices and aspects of material culture were adopted and did not, therefore, contribute to the emergence of a Creole-colonial empire in North America.

Notes

1 Rochefort, *Histoire naturelle et morale*, p. 15: 'Les Indiens qui n'avoient point encore veü de bâtiment de pareille figure, ni de matiere si solide, le consideroient au commencement avec un profond étonnement, et après avoir essayé avec la force de leurs épaules s'ils le pourroient ébranler, ils étoient contrains d'avoüer, que si toutes les maisons étoient bâties de la sorte, cette tempeste qu'on nomme *Ouragan*, ne les pourroit endommager.'

2 For an overview, see Edward E. Crain, *Historic Architecture in the Caribbean Islands* (Gainesville, FL: University Press of Florida, 1994).

3 Jean Davoigneau and Isabelle Duhau, 'Jacmel, entre rêve et réalité', *In Situ. Revue des patrimoines* 30 (2016), http://insitu.revues.org/13721, retrieved 4 June 2017; Maryse Sauphanor, *Maison creole* (Paris: Les auteur indépendantes, 2004); a very helpful inventory of the architectural heritage of Martinique is Jean-Luc Flohic, *Le patrimoine des communes de la Martinique*, 2nd ed. (Paris: Attique Editions, 2016); see also Danielle Bégot, *Atlas historique du patrimoine sucrier de la Martinique* (Paris: L'Harmattan, 1990); David Buisseret, *Histoire de l'architecture dans la Caraïbe* (Paris: Éditions caribéennes, 1984).

4 Most prominent is the case of Saint-Pierre on Martinique, where the early modern city was nearly completely wiped out by a pyroclastic flow caused by the volcanic eruption of Mount Pelée in 1902. Some older architecture remains on Haiti, but some have been damaged by the earthquake in 2010.

5 Aimé Césaire, 'Cahier d'un retour au pays natal', *Volonté* 20 (1939), new ed. (Paris/Dakar: Editions Présence africaine, 1993); I used the bilingual edition *The Original 1939 Notebook of a Return to the Native Land*, trans. and ed. A. James Arnold and Clayton Eshleman (Middletown, CT: Wesleyan University Press, 2013); the connection between Delawarde's photograph and Aimé Césaire's poem is made by Yves Bergeret, 'La case, ancrage symbolique du Cahier d'un retour au pays natal d'Aimé Césaire', in *Histoire par l'image*, www.histoire-image.org/etudes/case-ancrage-symbolique-cahier-retour-pays-natal-aime-cesaire, retrieved 5 June 2017.

6 Crain, *Historic Architecture*, p. 79.

7 Charlery, 'Maisons de maître et habitations coloniales'.

8 Préfontaine, *Maison rustique*, pp. 9–10.

9 For a reproduction of a detail of the plan, see Charlery, 'Maisons de maître et habitations coloniales', fig. 31.

10 Du Tertre, *Histoire générale des Antilles*, vol. 2, p. 450ff.: 'Les maisons des Gouverneurs sont toutes de pierre de taille et de moëllons. L'architecture en est assez reguliere, les chambres fort commodes, et à l'exterieur elles n'ont rien qui les distingue des bastiment en France.' Du Tertre himself makes the distinction of public and private buildings ('Des Bastimens, tant public que particulier') that indicates the importance of architecture for representing the state against the interests of the private landowners.

11 Labat, *Nouveau Voyage*, vol. 4, pp. 210–11: 'L'on voit assez par ce que je viens de dire, que J'avois disposé ce Bâtiment d'une maniere à pouvoir laisser tout l'étage

du rez de chausée à quelques Officier de consideration comme un Vice-Amiral de France, ou un Lieutenant General qui voudroient prendre leur logement à terre pendant leur séjour à la Martinique, sans que cela nous incommodant le moins du monde. Ainsi on pouvoit considerer le rez de chaussée comme une Maison seculiere, et le dessus comme un Couvent.'

12 CAOM 15DFC/812C–818C, 823A, 824C–829C, 839A, 831C–832C.
13 See Moreau de Saint-Méry, *Description topographique*, vol. 2, p. 625.
14 CAOM COL C8B 2, No. 69: La Broue to Pontchartrain, 21 February 1703: 'Les mils pieds de pierre de taille qui ont esté remis a Saint-Louis ne sont pas la, dixieme partie de celles dont il a besoint, ils l'ont cependant mis en estat de travailler à la porte du fort, et par le moyen qu'il a trouvé d'entirer d'une carriere esloignée de six lieues, Il pourra l'achever et faire le cordon des fortiffications, et les guerites. Il envoye les plans de l'estat des ouvrages de l'isle, on verra que les murailles sont eslevées de trois pieds de haut sur le rez de chaussée, et le petit front de 14, en sorte qu'on est a couvert d'un coup de main, il espere estre dans six mois au cordon, et avoir fii dans un an la fortiffication, s'il a 100 negres de plus [marginal remark by Pontchartrain's hand: "bon greffer"]. Il y a marqué les sousterreains qui sont estimez tres outils par tous les officiers. La place marquée de noir aus dessus de la porte est pour mettre une pierre de marbre sur la quelle sera gravée l'inscription qu'il plaira au Roy. Il sera necessaire de l'envoyer a un sculteur pour faire les ornemens. Il a demandé six mortiers qui seroient necessaires. Il n'est venu que six canons de 36 et 24, et il en faudroit encore 24 de 36 place a besoin de grosse artillerie, on peut placer jusques a 70 pieces de cette espece [marginal remark by Pontchartrain's hand: "bon, peu à peu"] le petite ne pouvant servir que pour les saluts.'
15 Zélie Navarro-Andraud, 'La résidence urbaine des administrateurs coloniaux de Saint-Domingue dans la seconde moitié du XVIIIe siècle', *Articulo. Journal of Urban Research*, Special Issue 1: Occupying, Organising and Ordering Urban Space (2009), http://articulo.revues.org/997, retrieved 7 June 2017, based on her thesis, 'Les élites urbaines de Saint-Domingue dans la seconde moitié du 18e siècle: La place des administrateurs coloniaux (1763–1792)' (PhD Dissertation, Université de Toulouse II-Le Mirail, 2007). Her argument is that colonial planning of Port-au-Prince as an administrative capital of the island led to a dysfunctional urban structure that only served as a whistle stop for goverment officials, traders, and local plantation owners leaving large areas of the town deserted. The idea seems very attractive in light of current efforts to rebuild the city centre of Port-au-Prince after its destruction by the earthquake in 2010. It is, however, problematic to assume from the perspective of a modern urban developer that the early modern government was able to realize their ideas and dreams without heavy compromises.
16 James E. McClellan, *Colonialism and Science: Saint Domingue in the Old Regime* (Baltimore, MD/London: Johns Hopkins University Press, 1992), pp. 75–108.
17 Trevor Burnard and John Garrigus, *The Plantation Machine: Atlantic Capitalism in French Saint-Domingue and British Jamaica* (Philadelphia, PA: University of Pennsylvania Press, 2016), p. 50.
18 CAOM 15DFC/5, No. 617 and 618.
19 On the buiding of water fountains on Guadeloupe, see Anne Pérotin-Dumon, *La ville aux Iles, la ville dans l'île: Basse-Terre et Pointe-à-Pitre, 1650–1820* (Paris: Karthala, 2000), pp. 371–9.
20 McClellan, *Colonialism and Science*, p. 88; Georges Corvington, *Port-au-Prince au cours des ans: La ville colonial, 1743–1789* (Port-au-Prince: Imprimerie Henri Deschampes, 1970).
21 Moreau de Saint-Méry, *Description topographique*, vol. 2, pp. 392–4.
22 Ibid., pp. 394–5.
23 Ibid., p. 395.
24 Ibid., p. 398.
25 Ibid.
26 Burnard and Garrigus, *Plantation Machine*, p. 55.

27 Moreau de Saint-Méry, *Description topographique*, vol. 1, p. 106: 'Le luxe y a donc un culte très suivi et c'est du Cap, comme d'un centre, qu'il répand ses jouissances et ses maux.'
28 Cited in Pluchon, *Histoire de la colonisation*, vol. 1, p. 387.
29 Charlery, 'Maisons de maître et habitations coloniales'.
30 Michel Étienne Descourtilz, *Voyages d'un naturaliste, et ses observations faites sur les trois règnes de la nature, dans plusieurs ports de mer français, en Espagne, au continent de l'Amérique Septentrionale, à Saint Yago de Cub*, 3 vols (Paris: Dufart, 1809), vol. 2, pp. 80–1: 'Nous entrâmes dans la ville du Cap, et la traversâmes pour arriver à nôtre logement qui étoit à l'extrémité opposée. Les colons, qui n'ont point quitté leur île, osent comparer cette capitale à Paris; mais ces deux cités ne souffrent point de parallèle. Les bâtiments du Cap étoient, à cette époque, construit sans goût, les rues étroites et horriblement pavées. Cependant cette ville, quoiqu'encore ensevelie sous les débris du pillage et de la dévastation, est encore le Paris de Saint-Domingue pour les ressources en tout genre qu'on y rencontre, et les ouvriers qui y sont, comme ailleurs, aux volontés de l'opulent.'
31 M. Desdorides, Remarques sur la colonie de St. Domingue, 1779, BNF Paris, Ms. 3453, cited in Pluchon, *Histoire de la colonisation*, vol. 1, p. 388.
32 See Pluchon, *Histoire de la colonisation*, vol. 1, pp. 388–9.
33 Pompée Valentin de Vastey, *An Essay on the Causes of the Revolution and Civil Wars of Hayti, Being a Sequel to the Political Remarks upon Certain French Publications and Journals Concerning Hayti*, trans. W. H. M. B. (Exeter: Western Luminary Office, 1823), p. 137.
34 Adrian Carton, *Mixed-Race and Modernity in Colonial India: Changing Concepts of Hybridity across Empires* (Abingdon/New York: Routledge, 2012), pp. 63–4.

Conclusion: The empire as a material construct

The struggle between local colonial interests and metropolitan imperial control, bearing at least some resemblance to the independence discourse in the wake of the American Revolution, was an exception, and emerged in France only late in the eighteenth century. A memorandum, printed in London in 1784, concerning the pros and cons of ceding parts or none of the commerce between the colonies and metropolitan France, shows the conflict was basically restricted to commercial proprietary or state interests arising from the institution of the *Exclusif* (trade restrictions).[1] The author of this anonymous publication was Jean-Baptiste Du Buc, otherwise known as the Grand Dubuc, of the extremely wealthy dynasty of planters from La Trinité in Martinique. He served as president of the island's powerful Chamber of Commerce and in 1761 naval secretary Choiseul chose him as his top administrator (*premier commis de la Marine*). His intention was to devise a mitigated version of the trade restriction (*exclusif mitigé*) that relaxed the laws prohibiting trade between France's colonies and other nations, in particular the United States of America.[2] One of his arguments in particular provides insight into the colonial mentality which regarded the colony not so much as a result of metropolitan initiative, but rather crediting its formation to those who had created a material 'culture':

1st proposition: The colonies have been created by the Metropolis and for the Metropolis.

Response: This assertion has two parts. The first is not accurate. A colony is not a summation of land, but a summation of culture. This culture is in the hand of the Negroes, and foreign nations and the French have procured these Negroes. Our merchants themselves have not stopped to denounce this foreign procuration; everybody knows that Guadeloupe in particular owes at least half of its culture, today so splendid, to the prodigious quantity of Negroes imported to this colony by the English, particularly in 1761, 1762 and 1763. Everybody knows that adventurers who had no relation to the commerce of France have established Saint-Domingue.[3]

The passage is remarkable insofar as by attempting to prove the dependence of the French colony on other European nations the author credits – certainly unintended by him – solely Africans for building the colony and thus laying the foundations for the marvellous prosperity of the island's economy. In fact, Du Buc articulates an all too obvious truth about African slaves being the prime agents for the creation and maintenance not only of the agriculture and the sugar industry indispensable for the Atlantic economy, but also of material culture in the wider sense of the word. 'Culture' is thus a cipher not only for the colony, its agriculture, and commerce, but includes the formation of empire, too. For empire as a formation includes the whole of the spatial production on the island and is thus much more than colonial commerce. Fortresses, public works, mansions, cottages, and even cities are the monumental manifestations of this culture that constitutes this particular notion of empire.

In theory, this conception of empire contradicts the standard model of empire where a centre exercises power and control over a periphery. At least for the French colonial realm, the distinction between an active metropolitan agency and a passive peripheral agency does not apply in all cases. The Antillean islands were colonies – these possessions were most often, in fact, called 'colonies' in the written and printed sources – that were ruled by a class of proprietors and administrative officials, but they were dependent more than all other French domains on the institution of African slavery.

The idea of continuing a Roman tradition of empire was not articulated by French officials during the seventeenth or eighteenth centuries. The historiographical standard model of a centre–periphery constellation only fits the Antillean case with regard to the economic and political framing of the islands as enclosed spaces where slavery was institutionalized and where property rights were granted priority in the political constitution of the colonies. But concerning the social interaction within this colonial framing, the reach of central authority was limited. The microhistorical study of large building projects on Martinique, Guadeloupe, and Saint-Domingue has shown that Africans, and in due course African-Atlantic Creoles, were responsible for constructing the material symbols and manifestations of colonial power. In the French Antilles both colonial power and Creole agency contributed to the realization of empire.

The French Antilles, therefore, adhered neither to a model of empire in the Roman tradition of having a strong centre, nor in the Athenian tradition of being a polity without a centre. A cosmopolitan commitment was as absent in France and its colonial possessions as were common ideas such as reason, liberty, or civilization. Only later, in the republican colonial empire of the late nineteenth and early twentieth century did such a justification narrative exist.[4] The first explicit reference to Athens and a new French empire of 'independent colonies' issued by the colonial reformer Charles-Maurice de Talleyrand must be seen in the context of the revolution and the crisis arising in the sugar-producing colonies of the Caribbean. But Talleyrand compares the lost colonies of Louisiana and Canada to the successful history of Greek colonies, since 'the colonists of those two countries were Frenchmen; they are so still; and an obvious bias inclines them always towards us'. The 'reciprocal interests', the masterful politician of this period of transition argues, 'strengthened by a powerful tie of common origin' are decisive for the coherence of the new empire.[5]

But clearly, the Antillean colonies were not independent in this sense and therefore no good example for the renewal of the empire of the Ancien Régime. Quite to the contrary Talleyrand pleaded in favour of disbanding the idea of production colonies like Saint-Domingue and establishing a new system of plantation economies in West Africa 'and in those places where the cultivator is born'.[6] Apart from being too optimistic about this plan of production colonies in Africa – an idea that had already been mentioned in the seventeenth century by the director Chambonneau[7] – the French did indeed fail to foster 'reciprocal interests' in the Antilles. The 'ideal' French colony they were able to build in Louisbourg, for example, was very different to the islands, where there was no 'system of interaction between two political entities' that could be called, according to Doyle, 'effective sovereignty'.[8]

Instead, the Antillean type of empire was rather similar to that of the Greco-Roman world, consisting mostly of 'permanent exploitation' and gain.[9] In this sense, the empire of the Antilles was rather unintended and a result of accidental circumstance. From an economic perspective, the French empire in the Caribbean was a result of the actions of certain groups interested in the process of automated empire formation that were not restricted to a centre, but could also have been at the supposed periphery. Therefore, the class system has been perhaps more dominant in the Antilles than in other parts of the colonial realm.[10] In this respect, the French Antilles can be compared to the Dutch empire, in which many different groups from all over the Indo-Pacific realm participated in a system of exploitation and gain.

The dynamics of the practices of material empire building in the Antilles uncover another aspect that contributed to the formation of a polity that might seem cosmopolitan after all. The participation of a multitude of actors in the construction of the built environment might have furthered a cosmopolitan practice. Did the enslaved builders contribute to the formation of an empire in an oppositionary capacity in order to build a 'counter-empire' in the sense of Hardt and Negri, or were they actually included, or rather entangled, in an open and spatially variable structure of empire in the sense of Beck and Grande?[11] Both notions of cosmopolitan empire appear applicable to the Antillean case only insofar as they underline the supranational or 'hybrid' character of different institutions empowering each other that was so typical for the early modern period. However, on closer examination there are problems with this model, too: Hardt's and Negri's discursive super-structure of an empire that 'manages hybrid identities, flexible hierarchies, and plural exchanges' etc. is not intentionally managed by a regime that enforces some kind of order.[12] Also, Beck's and Grande's concept of empire presupposes a practice of 'strategically binding others', exercised by institutions that did not exist on the Antillean islands.[13]

There are, however, alternative functions of a cosmopolity in empire that transgress the conceptional idealizations that I have previously discussed in respect of the philosophical models of empire. The restriction to only one place in the French colonial realm offers insights into the inner dynamics of a society formed by colonial and Creole agency. Existing models of empire – those presupposing a centre and those that do not – have shown limitations that can only be overcome if one widens the scope to include comparable and globally entangled locales. The formulation of a relational model of empire should help to collect together the different threads of practice under variable spatial conditions and to weave a new narrative fabric of the French empire.

Four threads, with respect to four different places, within this fabric were studied in the previous chapters. The outcome of this large microhistory has been a picture of a variegated empire. Not so much an empire of fringes – that presupposes a fabric of power and control more closely knit towards the centre – the French realm seems to have been, to take the metaphor further, more of a rag-rug empire. Each of these locations appears to have a particular character: a Creole-colonial or mixed society in the Antilles and Senegambia, accommodation to an Indian idea of empire in Pondichéry, and a colonial culture in Atlantic Canada. In this sense the four cases seem to have been individual instances of imperial implementation practices. And since there was no imperial centre resourceful enough to have devised these practices

by intent, the individual colonial spaces appear to have been discon-
nected, fragmented, and virtually independent of metropolitan France
and each other.

But a large microhistory does not restrict itself to a comparative
history that marks differences and the singularity of individual cases.
It is a connected history, too. The imperial fabric was made up of single
pieces, but spatially connected by means of material, emotions, and
style. There were similarities between the building practices, the logistics,
the co-operation with local contractors, the introduction of a classical
Ludovican style, symbolic ornaments, and the methods of territorial
enclosure. The space of the French empire can, therefore, be conceived
as a non-linear expanse with connections that link different places
closely together. The picture would be that of a tubulous world where
the different colonial zones are connected by epistemic, aesthetic, and
material correspondence.

Returning to the example of Haiti mentioned at the beginning of
the book, it may be asked what the connection between France and
the former colony of Saint-Domingue looked like. The most important
practice of empire building was, as argued throughout the book, hopefully
conclusively, the transformation of the natural environment of islands,
peninsulas, coastlines, rivers, and mountains into an enclosed territory.
This was done not only by drawing lines on maps and thereby creating
a representation of territorial space on paper, but also by physically
planting these lines as borders into the earth. Islands were turned into
colonies, sometimes, as in the case of Saint-Domingue, divided by a
mountainous border in the middle of the island. This practice of enclosure
brought about a colonial space that lasted even after the French had
to leave the new independent nation of Haiti.

But the practice was merely the framing of an imperial space that
allowed for all sorts of different local, indigenous, and Creole agencies
and practices to unfold and participate in the configuration of the place.
In this respect, empire became a living polity, a 'cultural practice'
(Sheldon Pollock) that was able to undermine the colonial order established by the territorial transformation, or else further it by adapting
to its spatial possibilities, even contributing to its continuation and
improvement for different governments and commercial elites.

It is not easy to admit that empire was in this respect – and perhaps
particularly from the perspective of large building projects – a more
inclusive than exclusive process in history. But the aim of this book
is not to present empire as a utopian vision of a harmonious community
that embraced all its members. Even if the people of this empire were
described as an emotional community, one must keep in mind that
these emotions were those of suffering, oppression, and trauma. Many

of the representative buildings with a distinct colonial style are reminders of the unfathomable tragedy of slavery and the exploitation of so many people, whose descendents are still witnessing its consequences. The empire is thus ultimately a figure in memory as much as it is a feature of history. It stands for a material figuration that must be understood in all its different facets in order to explain the historical foundation of modern societies in France and its empire.

All the different locales chosen as the objects of this inquiry – the Antilles, Senegambia, Pondichéry, Louisbourg – have their respective memory of this period of imperial history. Some colonial spaces were reoccupied by their inhabitants, liberated in Haiti, decolonized in Pondichéry, amalgamated in Saint-Louis, simply deteriorated as in the case of Louisbourg, or served as a national narrative of exceptionalism as in Québec. In France, however, the memory of this deep history of empire is ambivalent. Its present society has yet come to terms with its global involvement with other cultures and people, to whom it is indebted both in a positive sense in creating the opportunities for exchange, integration, and even splendour, but also in a negative sense in creating injustices, suffering, and the hardships of mostly unrewarded labour.

Notes

1   CAOM 07DFC/127, No. 91: [Jean-Baptiste Du Buc]: *Le Pour et le Contre sur un objet de grande discorde et d'importance majeure. Convient-il à l'Administration de céder part, ou de ne rien céder aux Etrangers dans le Commerce de la Métropole avec ses Colonies?* (London 1784).

2   See Jean Tarrade, *Le commerce colonial de France à la fin de l'Ancien Régime: l'évolution du régime de l'exclusif de 1763 à 1789*, 2 vols (Paris: Presses universitaires de France, 1972), vol. 1, pp. 185, 190–211.

3   CAOM 07DFC/127, No. 91: [Jean-Baptiste Du Buc]: *Le Pour et le Contre*, p. 1f.: '1°. Les Colonies ont été créées par la Métropole et pour la Métropole. – R. Cette assertion a deux parties. La premiere n'est pas exacte. Une Colonie n'est pas une somme de terre, mais une somme de culture. Cette culture est dans la main des Noirs, et ces Noirs ont été fournis par les Etrangers et par les Français. Nos Négociants eux-mêmes n'ont cessé de denouncer cette fourniture étrangere; tout le monde sait que la Guadeloupe en particulier doit au moins la moitié de sa culture, aujourd'hui si brillante, à la prodigieuse quantité des Négres importés dans cette Colonie par les Anglais, sur-tout un 1761, 1762 et 1763. Tout le monde sait que Saint-Domingue a été établi par des Aventuriers qui n'avaient aucune relation avec le Commerce de France.'

4   See Alice L. Conklin, *A Mission to Civilize: The Republican Idea of Empire in France and West Africa, 1895–1930* (Stanford, CA: Stanford University Press, 1997); Boris Barth and Jürgen Osterhammel (eds), *Zivilisierungsmissionen. Imperiale Weltverbesserung seit dem 18. Jahrhundert*, Historische Kulturwissenschaften, 6 (Konstanz: UVK, 2005); Dino Costantini, *Mission civilisatrice. Le rôle de l'histoire coloniale dans la construction de l'identité politique française* (Paris: La Découverte, 2008).

5 Charles-Maurice de Talleyrand, *Essai sur les avantages à retirer de colonies nouvelles dans les circonstance présentes, par le citoyen Talleyrand. Lu à la séance publique de l'Institut national le 15 méssidor an 5* [1797]; published in English as 'An Essay on the Advantages to Be Derived from New Colonies, in the Present Circumstances', *The Colonial Journal* 4 (1816), 322–8.

6 Ibid., p. 328.

7 CAOM Col C6 1: Chambonneau to Seignelay, July 1688, fol. 2v. See also Benjamin Steiner, *Colberts Afrika. Eine Wissens- und Begegnungsgeschichte in Afrika im Zeitalter Ludwigs XIV.* (München: Oldenbourg, 2014), p. 373; Cultru, *Histoire du Sénégale*, p. 118.

8 Doyle, *Empires*, p. 12.

9 Moses Finley, 'Empire in the Greco-Roman World', *Greece & Rome* 25:1 (1978), 6. According to Finley, 'there is no empire' without 'permanent exploitation'. In his argument on the unintentional formation of the Athenian and Republican Roman Empire, however, he contends that the profiteers of these empires did not only reside in the administrative centre, but that a multitude of peripheral historical actors, including local aristocrats in the Roman Empire and lower classes in the Athenian democratic empire, could be counted among those who benefitted by such a political and economic order.

10 For the issue of class, see Vera S. Candiani, 'The Desagüe Reconsidered: Environmental Dimensions of Class Conflict in Colonial Mexico', *Hispanic American Historical Review* 92:1 (2012), 5–26.

11 Michael Hardt and Antonio Negri: *Empire* (Cambridge, MA: Harvard University Press, 2000); further elaborated in *Multitude: War and Democracy in the Age of Empire* (London: Penguin, 2004), and *Commonwealth* (Cambridge, MA: Harvard University Press, 2009); for a critical reception, see Heinz-Dieter Kittsteiner, 'Empire. Zu den revolutionären Phantasien von Antonoio Negri und Michael Hardt', in Richard Faber (ed.), *Imperialismus in Geschichte und Gegenwart* (Würzburg: Königshausen und Neumann, 2005), pp. 125–50.

12 Hardt and Negri, *Empire*, pp. xii–xiii.

13 Ulrich Beck and Edgar Grande, *Cosmopolitian Europe*, trans. Ciaran Cronin (Cambridge: Polity, 2007), p. 81.

BIBLIOGRAPHY

Archival sources

Aix-en-Provence
Centre des archives d'outre-mer (CAOM)
  Fonds concernant l'outre-mer
  Colonies. Archives ministérielles anciennes (COL)
  Série C (Correspondance de l'arrivée)
    C 6 (Senegal): 1
    C 8A (Martinique): 1, 2, 3, 5, 6, 21, 24, 25, 41, 42, 44, 50, 57
    C 8B (Martinique): 1, 2, 6, 22
  Dépot des fortifications des colonies (DFC)
    03DFC (Port Royale, Acadie): 6
    07DFC (Antilles): 127
    08DFC (Guadeloupe): 26, 27
    13DFC (Martinique): 49, 50
    15DFC (Saint-Domingue):1, 5
    16DFC (Côte d'Afrique, Senegal): 74, 82
    17DFC (Senegal): 76, 77
    26DFC (Pondichéry): 98, 99

Published sources

Adelman, Jeremy, and Steven Aron, 'From Borderlands to Borders: Empires, Nation States and Peoples in between the North American History', *The American Historical Review* 104 (1999), 814–41.

Agamben, Giorgio, *Homo Sacer: Sovereign Power and Bare Life* (Stanford, CA: Stanford University Press, 1988).

Alam, Muzaffar, and Sanjay Subrahmanyam (eds), *The Mughal State* (Oxford: Oxford University Press, 1997).

Allaire, Louis, 'The Caribs of the Lesser Antilles', in Samuel M. Wilson (ed.), *The Indigenous People of the Caribbean* (Gainesville, FL: University of Florida Press, 1997), pp. 180–5.

Alpers, Svetlana, 'Style Is What You Make It: The Visual Arts Once Again', in Berel Lang (ed.), *The Concept of Style*, revised and expanded ed. (Ithaca, NY/London: Cornell University Press, 1987), pp. 138–62.

Ames, Glenn J., *Colbert, Mercantilism, and the French Quest for Asian Trade* (DeKalb, IL: Northern Illinois University Press, 1996).

Amselle, Jean-Loup, *Mestizio Logics: Anthropology of Identity in Africa and Elsewhere*, translated by Claudia Royal (Stanford, CA: Stanford University Press, 1998).

Anderson, Fred, *Crucible of War: The Seven Years' War and the Fate of Empire in British North America, 1754–1766* (London: Vintage, 2001).

Andrade, Tonio, *How Taiwan Became Chinese: Dutch, Spanish, and Han Colonization in the Seventeenth Century* (New York: Columbia University Press, 2007).

Andrade, Tonio, 'A Chinese Farmer, Two Black Boys, and a Warlord: Towards a Global Microhistory', *The Journal of World History* 21:4 (2011), 573–91.

Andrade, Tonio, and William Reger (eds), *The Limits of Empire: European Imperial Formations in Early Modern World History. Essays in Honor of Geoffrey Parker* (Farnham/Burlington, VA: Ashgate, 2012).

Andrien, Kenneth J., 'The Spanish Atlantic System', in Jack P. Greene and Philip D. Morgan (eds), *Atlantic History: A Critical Appraisal* (Oxford: Oxford University Press, 2009), pp. 55–81.

Appadurai, Arjun, 'Introduction: Commodities and the Politics of Value', in Arjun Appadurai (ed.), *The Social Life of Things: Commodities in Cultural Perspective* (Cambridge: Cambridge University Press, 1986), pp. 3–63.

Armitage, David, 'The Scottish Vision of Empire: Intellectual Origins of the Darien Venture', in John Robertson (ed.), *A Union for Empire: Political Thought and the British Union of 1707* (Cambridge: Cambridge University Press, 1995), pp. 97–118.

Armitage, David, 'Making the Empire British: Scotland in the Atlantic World, 1542–1707', *Past and Present* 155 (1997), 34–63.

Armitage, David, *The Ideological Origins of the British Empire* (Cambridge: Cambridge University Press, 1999).

Armitage, David, and Michael J. Braddick (eds), *The British Atlantic World, 1500–1800* (London/New York: Palgrave Macmillan, 2002).

Armitage, David, and Jo Guldi, *Historical Manifesto* (Cambridge: Cambridge University Press, 2014).

Asher, Catherine B., *The New Cambridge History of India*, Part I, vol. 4: Architecture of Mughal India (Cambridge: Cambridge University Press, 1992).

Bailey, Gauvin Alexander, *Architecture and Urbanism in the French Atlantic Empire: State, Church, and Society, 1604–1830*, McGill-Queen's French Atlantic Worlds Series, 1 (Toronto: McGill-Queen's University Press, 2018).

Bailey, Gauvin Alexander, *Colonial Architecture Project*, www.colonialarchitectureproject.org, retrieved 13 June 2018.

Bailyn, Bernard, *The Ideological Origins of the American Revolution* (Cambridge, MA: Belknap Press, 1992).

Bailyn, Bernard, *Atlantic History: Concepts and Contours* (Cambridge, MA: Harvard University Press, 2005).

Balcom, B. A., *The Cod Fishery of Isle Royale, 1713–58* (Ottawa: National Historic Parks and Site Branch Parks Canada, 1984).

Balcom, B. A., 'Defending Unama'ki: Mi'kmaw Resistance in Cape Breton, 1745', *Nashwaak Review* 22–23 (2009), 447–92.

Balesi, Charles John, *The Time of the French in the Heart of North America: 1673–1818* (Chicago, IL: Alliance française, 1992).

Ballantyne, Tony, *Webs of Empire: Locating New Zealand's Colonial Past* (Wellington: Bridget Williams Books, 2012).

Banbuck, Cabuzel-Andréa, *Histoire politique, économique et sociale de la Martinique sous l'ancien régime (1635–1789)* (Paris: Librairie des sciences politiques et sociales, 1935; repr. Fort-de-France: Société de distribution et de culture, 1972).

Banks, Kenneth, '"Lente et assez fâcheuse traverse": Navigation and the Trans-atlantic French Empire, 1713–1763', in A. J. B. Johnston (ed.), *Proceedings of the Twentieth Meeting of the French Colonial Historical Society* (Cleveland: FCHS, 1994), pp. 80–94.

Banks, Kenneth, *Chasing Empire across the Sea: Communication and the State in the French Atlantic, 1713–1763* (Montreal/London/Ithaca: McGill-Queen's University Press, 2003).

Banks, Kenneth, 'Communications and "Imperial Overstrech": Lessons from the Eighteenth-Century French Atlantic', *French Colonial History* 6 (2005), 17–32.

Banks, Kenneth, 'Financiers, Factors, and French Proprietary Companies in West Africa, 1664–1713', in L. H. Roper and B. Van Ruymbeke (eds), *Constructing Early Modern Empires: Proprietary Ventures in the Atlantic World, 1500–1750* (Leiden/Boston, TX: Brill, 2007), pp. 79–116.

Barad, Karen, *Meeting the Universe Halfway: Quantum Physics and the Entanglement of Matter and Meaning* (Durham: Duke University Press, 2007).

Barth, Boris, and Jürgen Osterhammel (eds), *Zivilisierungsmissionen. Imperiale Weltverbesserung seit dem 18. Jahrhundert*, Historische Kulturwissenschaften, 6 (Konstanz: UVK, 2005).

Bayly, Christopher A., *Imperial Meridian: The British Empire and the World, 1780–1830* (Harlow: Longman, 1989).

Bayly, Christopher A., *Empire and Information: Intelligence Gathering and Social Communication in India, 1780–1870* (Cambridge: Cambridge University Press, 1996).

Bayly, Christopher A., *The Birth of the Modern World, 1780–1914* (Malden, MA/Oxford: Blackwell, 2004).

Beck, Ulrich, *Weltrisikogesellschaft* (Frankfurt am Main: Suhrkamp, 2007).

Beck, Ulrich, and Edgar Grande, *Cosmopolitan Europe*, translated by Ciaran Cronin (Cambridge: Polity, 2007).

Bégot, Danielle, *Atlas historique du patrimoine sucrier de la Martinique* (Paris: L'Harmattan, 1990).

Bellin, Jacques-Nicolas, *Carte réduite des isles de la Guadeloupe, Marie-Galante et les Saintes* (Paris: Dépôt des cartes plans et journaux de la marine, 1759).

Benton, Lauren, *A Search of Sovereignty: Law and Geography in European Empires, 1400–1900* (Cambridge: Cambridge University Press, 2010).

Bergeret, Yves, 'La case, ancrage symbolique du Cahier d'un retour au pays natal d'Aimé Césaire', in *Histoire par l'image*, www.histoire-image.org/etudes/case-ancrage-symbolique-cahier-retour-pays-natal-aime-cesaire, retrieved 5 June 2017.

Bermingham, Ann, and John Brewer (eds), *The Consumption of Culture, 1600–1800: Image, Object, Text* (London: Routledge, 1995).

Bertrand, Romain, and Guillaume Calafat, 'La microhistoire globale: affaire(s) à suivres', *Annales: Histoire, Sciences Sociales* 73 (2018), 3–19.

Bitterling, David, 'Marschall Vauban und die absolute Raumvorstellung', in Lars Behrisch (ed.), *Vermessen, Zählen, Berechnen: Die politische Ordnung des Raums im 18. Jahrhundert* (Frankfurt am Main/New York: Campus, 2006), pp. 65–74.

Bitterling, David, *L'invention du pré carré. Construction de l'espace français sous l'Ancien Régime* (Paris: Albin Michel, 2009).

Black, Jeremy, *The British Empire: A History and a Debate* (Burlington: Ashgate, 2015).

Blackburn, Robin, *The Making of New World Slavery: From the Baroque to the Modern, 1492–1800* (New York/London: Verso, 1998).

Blanchard, Anne, *Les ingénieurs du 'roy' de Louis XIV à Louis XVI. Étude du corps des fortifications* (Montpellier: Déhan, 1979).

Bleichmar, Daniela, *Visible Empire: Botanical Expeditions and Visual Culture in the Hispanic Enlightenment* (Chicago, IL: University of Chicago Press, 2012).

Bleichmar, Daniela, and Peter C. Mancall (eds), *Collecting across Cultures: Material Exchanges in the Early Modern Atlantic World* (Philadelphia, PA: University of Pennsylvania Press, 2011).

Boucher, Philip P., *The Shaping of the French Colonial Empire: Bio-bibliography of the Careers of Richelieu, Fouquet, and Colbert* (New York: Garland Press, 1985).

Boucher, Philip P., *Cannibal Encounters: Europeans and Island Caribs, 1492–1763* (Baltimore, MD/London: Johns Hopkins University Press, 1992).

Boucher, Philip P., 'The "Frontier Era" of the French Caribbean, 1620s–1690s', in Christine Daniels and Michael V. Kennedy (eds), *Negotiated Empires: Centers and Peripheries in the Americas, 1500–1820* (New York/London: Routledge, 2002), pp. 207–35.

Boucher, Philip P., 'French Proprietary Colonies in the Greater Caribbean, 1620s–1670s', in L. H. Roper and B. Van Ruymbeke (eds), *Constructing Early Modern Empires: Proprietary Ventures in the Atlantic World, 1500–1750* (Leiden/Boston, TX: Brill, 2007), pp. 163–88.

Boucher, Philip P., *France and the American Tropics to 1700: Tropic of Discontent?* (Baltimore, MD: Johns Hopkins University Press, 2008).

Bourdieu, Pierre, *Outline of a Theory of Practice* (Cambridge: Cambridge University Press, [1972] 1977).

Bourguet, Marie-Noelle, Christian Licoppe, and H. Otto Sibum (eds), *Instruments, Travel, and Science: Itineraries of Precision from the Seventeenth to the Twentieth Century* (London/New York: Routledge, 2002).

Boxer, Charles R., *The Dutch Seaborne Empire, 1600–1800* (Oxford: Clarendon, 1965).

Boxer, Charles R., *The Portuguese Seaborne Empire, 1415–1825* (New York: Knopf, 1969).

Braudel, Fernand, *Civilisation matérielle et capitalisme. XVe – XVIIIe siècle*, 3 vols (Paris: Armand Colin, 1967–1979); published in English as *Civilization and Capitalism, 15th–18th Century*, 3 vols (London: Collins, 1982–4).

Brendecke, Arndt, *The Empirical Empire: Spanish Colonial Rule and the Politics of Knowledge* (Berlin: de Gruyter/Oldenbourg, 2016).

Bright, Alistair J., '"Removed from the Face of the Island": Late Pre-colonial and Early Colonial Amerindian Society in the Lesser Antilles', in Corinne Lisette Hofman and Anne van Duijvenbode (eds), *Communities in Contact:*

Essays in Archaeology, Ethnohistory & Ethnography of the Amerindian Circum-Caribbean (Leiden: Sidestone Press, 2011), pp. 307–25.

Brooks, George E., *Eurafricans in Western Africa: Commerce, Social Status, Gender, and Religious Observance from the Sixteenth to the Eighteenth Century* (Athens, OH: Ohio University Press, 1998).

Bolton, Herbert Eugene, 'The Epic of Greater America', *The American Historical Review* 38:3 (1933), 448–74.

Buisseret, David, *Histoire de l'architecture dans la Caraïbe* (Paris: Éditions caribéennes, 1984).

Buisseret, David, *Ingénieurs et fortifications avant Vauban. L'organisation d'un service royal aux XVIe–XVIIe siècles* (Paris: Éditions du C.T.H.S., 2002).

Buisseret, David, 'French Cartography: The *ingénieurs du roi*, 1500–1600', in David Woodward (ed.), *The History of Cartography* (Chicago, IL: University of Chicago Press, 2007), vol. 3, pp. 1504–21.

Burbank, Jane, and Frederick Cooper, *Empires in World History: Power and Politics of Difference* (Princeton, NJ: Princeton University Press, 2011).

Burnard, Trevor, 'The British Atlantic', in Jack P. Greene and Philip D. Morgan (eds), *Atlantic History: A Critical Appraisal* (Oxford: Oxford University Press, 2009), pp. 111–37.

Burnard, Trevor, and John Garrigus, *The Plantation Machine: Atlantic Capitalism in French Saint-Domingue and British Jamaica* (Philadelphia, PA: University of Pennsylvania Press, 2016).

Bushnell, Amy Turner, and Jack P. Greene, 'Peripheries, Centers, and the Construction of Early Modern American Empires: An Introduction', in Christine Daniels and Michael J. Kennedy (eds), *Negotiated Empires: Centers and Peripheries in the Americas, 1500–1820* (New York/London: Routledge, 2002), pp. 1–14.

Butel, Paul, *Histoire des Antilles françaises, XVIIe–XXe siècle* (Paris: Perrin, 2007).

Camus, Michel-Christian, 'Le général de Poincy, premier capitaliste sucrier des Antilles', *Revue française d'histoire d'outre-mer* 84 (1997), 119–25.

Candiani, Vera S., 'The Desagüe Reconsidered: Environmental Dimensions of Class Conflict in Colonial Mexico', *Hispanic American Historical Review* 92:1 (2012), 5–26.

Candiani, Vera S., *Dreaming of Dry Land: Environmental Transformation in Colonial Mexico City* (Stanford, CA: Stanford University Press, 2014).

Capek, Milic, *The Concepts of Space and Time: Their Structure and Their Development* (Dordrecht/Boston, TX: Reidel, 1976).

Cardim, Pedro, Tamar Herzog, Jose Javier Ruiz Ibáñez, and Gaetano Sabatini (eds), *Polycentric Monarchies: How Did Early Modern Spain and Portugal Achieve and Maintain a Global Hegemony?* (Brighton/Portland: Sussex Academic Press, 2012).

Carton, Adrian, *Mixed-Race and Modernity in Colonial India: Changing Concepts of Hybridity across Empires* (London/New York: Routledge, 2012).

de Certeau, Michel, *L'écriture de l'histoire* (Paris: Gallimard, 1975), pp. 219–48.

de Certeau, Michel, 'Pratique d'espace', in *L'invention du quotidian*, edited by Luce Giard, vol. 1: Arts de faire (Paris: Gallimard, 1990), pp. 139–91.

Césaire, Aimé, 'Cahier d'un retour au pays natal', *Volonté* 20 (1939), new edition (Paris/Dakar: Editions Présence africaine, 1993).

Césaire, Aimé, *The Original 1939 Notebook of a Return to the Native Land*, translated and edited by A. James Arnold and Clayton Eshleman (Middletown, CT: Wesleyan University Press, 2013).

Charlery, Christophe, 'Maisons de maître et habitations coloniales dans les anciens territoires français de l'Amérique tropicale', *In Situ. Revue des patrimoines* 5 (2004), http://insitu.revues.org/2362, retrieved 9 May 2017.

Charles, Loïc, and Paul Cheney, 'The Colonial Machine Dismantled: Knowledge and Empire in the French Atlantic', *Past & Present* 219 (2013), 127–63.

Chaunu, Pierre, and Hugette Chaunu, *Séville et l'Atlantique, 1504–1650*, 8 vols (Paris: A. Colin, 1955–60).

Chénier, Rémi, *Québec: A French Colonial Town in America, 1660 to 1690* (Ottawa: Minister of the Environment, 1991).

Choquette, Leslie, *Frenchmen into Peasants: Modernity and Tradition in the Peopling of French Canada* (Cambridge, MA/London: Harvard University Press, 1997).

Chouin, Gérard, *Colbert et la Guinée. Le voyage en Guinée de Louis de Hally et Louis Ancelin de Gémozac (1670–1671)* (Saint-Maur-des-Fossés: Sepia, 2011).

Collet, Dominik, *Die Welt in der Stube. Begegnungen mit Außereuropa in Kunstkammern der Frühen Neuzeit* (Göttingen: Vandenhoeck & Ruprecht, 2007).

Conklin, Alice L., *A Mission to Civilize: The Republican Idea of Empire in France and West Africa, 1895–1930* (Stanford, CA: Stanford University Press, 1997).

Cook, Shelburne F., and Woodrow Borah, *Essays in Population History: Mexico and the Caribbean* (Berkeley, CA: University of California Press, 1971).

Coole, Diana, and Samantha Frost (eds), *New Materialisms: Ontology, Agency, and Politics* (Durham/London: Duke University Press, 2010).

Cooper, Frederick, 'Empire Multiplied: A Review Essay', *Comparative Studies in Society and History* 46:2 (2004), 247–72.

Cormack, Lesley B., *Charting an Empire: Geography at the English Universities, 1580–1620* (Chicago, IL: University of Chicago Press, 1997).

Corvington, Georges, *Port-au-Prince au cours des ans: La ville colonial, 1743–1789* (Port-au-Prince: Imprimerie Henri Deschampes, 1970).

Cosgrove, Denis, *Apollo's Eye: A Cartographic Genealogy of the Earth in Western Imagination* (Baltimore, MD: Johns Hopkins University Press, 2001).

Cosgrove, Denis, *Geography and Vision: Seeing, Imagining and Representing the World*, International Library of Human Geography, 12 (London/New York: I. B. Tauris, 2008).

Costantini, Dino, *Mission civilisatrice. Le rôle de l'histoire coloniale dans la construction de l'identité politique française* (Paris: La Découverte, 2008).

Crain, Edward E., *Historic Architecture in the Caribbean Islands* (Gainesville, FL: University Press of Florida, 1994).

Cremer, Annette, and Martin Mulsow (eds), *Objekte als Quellen der historischen Kulturwissenschaften: Stand und Perspektiven der Forschung* (Köln, Weimar, Wien: Böhlau, 2018).

Crépin, Pierre, *Mahé de La Bourdonnais, gouverneur général des Îles de France et de Bourbon, 1699–1753* (Paris: Leroux, 1922).

Crouse, Nellis M., *French Pioneers in the West Indies, 1665–1713* (New York: Columbia University Press, 1940).

Cultru, Prosper, *Histoire du Sénégal du XVe siècle à 1870* (Paris: Emile Larose, 1910).

Cultru, Prosper, 'Colonisation d'autrefois. Le Commandeur de Poincy à Saint-Christophe', *Revue de l'histoire des colonies françaises* (1915), 289–354.

Curry, Ginette, *"Toubab La!" Literary Representations of Mixed-Race Characters in the African Diaspora* (Newcastle: Cambridge Scholars Publishing, 2007).

Curtin, Philip D., *The Atlantic Slave Trade: A Census* (Madison, WI: University of Wisconsin Press, 1969).

Dale, Ronald J., *The Fall of New France: How the French Lost a North American Empire, 1754–1763* (Toronto: Lorimer, 2004).

Daviler, Augustin Charles, *Cours d'architecture, qui comprend les ordres de Vignole* (Paris: Nicolas Langlois, 1691).

Davoigneau, Jean, and Isabelle Duhau, 'Jacmel, entre rêve et réalité', *In Situ. Revue des patrimoines* 30 (2016), http://insitu.revues.org/13721, retrieved 4 June 2017.

Dawdy, Shannon Lee, *Building the Devil's Empire: French Colonial New Orleans* (Chicago, IL: University of Chicago Press, 2009).

Debien, Gabriel, *Les Engagés pour les Antilles (1634–1715). La société coloniale aux 17e et 18e siècle* (Paris: Larose, 1952).

Debien, Gabriel, *Plantations et esclaves à Saint-Domingue* (Dakar: Université de Dakar, 1962).

Debien, Gabriel, *Les Grand'cases des plantations à Saint-Domingue aux XVIIe et XVIIIe siècles*, Annales des Antilles, Bulletin de la société d'histoire de la Martinique, 15 (Fort-de-France: Société d'histoire de la Martinique, 1970).

Debien, Gabriel, *Les esclaves aux Antilles Françaises (XVIIe–XVIIIe siècles)* (Basse-Terre/Fort-de-France: Société d'Histoire de la Guadeloupe/Société d'Histoire de la Martinique, 1974).

DeCorse, Christopher R., *An Archaeology of Elmina: Africans and Europeans on the Gold Coast, 1400–1900* (Washington, DC: Smithsonian Institution Press, 2001).

Delcourt, André, *La France et les établissements français au Sénégal, 1713–1769*, Mémoires de l'Institut français d'Afrique noire, 17 (Dakar: IFAN, 1952).

Deleuze, Gilles, amd Felix Guattari, *A Thousand Plateaus: Capitalism and Schizophrenia* (Minneapolis, MN: University of Minnesota Press, 1987).

Deloche, Jean (ed.), *Le papier terrier de la ville blanche de Pondichéry, 1777* (Pondichéry: Institut français de Pondichéry, 2002).

Deloche, Jean, *Le vieux Pondichéry (1673–1824) revisité d'après les plans anciens* (Pondichéry: Institut français de Pondichéry, 2005).

Deloche, Jean, *Pondicherry Past and Present: Pondichéry hier et aujourd'hui*, 2nd ed. (Pondichéry: Institut français de Pondichéry, 2019).

Descourtilz, Michel Étienne, *Voyages d'un naturaliste, et ses observations faites sur les trois règnes de la nature, dans plusieurs ports de mer français, en Espagne, au continent de l'Amérique Septentrionale, à Saint Yago de Cub*, 3 vols (Paris: Dufart, 1809).

Dessert, Daniel, *Argent, pouvoir et société au Grand Siècle* (Paris: Fayard, 1984).

Dew, Nicholas, 'Scientific Travel in the Atlantic World: The French Expedition to Gorée and the Antilles, 1681–1683', *The British Journal for the History of Science* 43:1 (2010), 1–17.

Diderot, Denis, and Jean Le Rond d'Alembert, 'Ingénieur', in *Encyclopédie, ou Dictionnaire raisonné des sciences, des arts et de métiers*, vol. 8 (Paris: Briasson, David, Le Breton, Durand, 1766), pp. 741–3.

Disney, A. R., *A History of Portugal and the Portuguese Empire*, 2 vols (Cambridge: Cambridge University Press, 2009).

Donavan, Kenneth, 'Slaves and Their Owners in Ile Royale, 1713–1760', *Acadiensis* 25:1 (1995), 3–32.

Douglas, Mary, and Barry Isherwood, *The World of Goods: Towards an Anthropology of Consumption* (Harmondsworth: Penguin, 1978).

Doyle, Michael, *Empires* (Ithaca, NY/London: Cornell University Press, 1984).

Drayton, Richard H., 'Knowledge and Empire', in P. J. Marshall (ed.), *The Oxford History of the British Empire*, vol. 2: The Eighteenth Century (Oxford: Oxford University Press, 1998), pp. 231–52.

Dubois, Laurent, 'The French Atlantic', in Jack P. Greene and Philip D. Morgan (eds), *Atlantic History: A Critical Appraisal* (Oxford: Oxford University Press, 2009), pp. 137–61.

Dubois, Laurent, 'Slavery in the French Caribbean, 1635–1804', in David Eltis and Stanley L. Engerman (eds), *The Cambridge World History of Slavery*, vol. 3: 1420–1804 (Cambridge: Cambridge University Press, 2011), pp. 431–49.

DuVal, Kathleen, *The Native Ground: Indians and Colonists in the Heart of the Continent* (Philadelphia, PA: University of Pennsylvania Press, 2007).

Eccles, W. J., *Canada under Louis XIV, 1663–1701* (Oxford/New York: Oxford University Press, 1964).

Einstein, Albert, 'Das Raum-, Äther- und Feld-Problem der Physik', in *Mein Weltbild*, edited by Carl Seelig (Zürich/Stuttgart/Wien: Europa-Verlag, 1953), pp. 181–93.

Eisenstadt, Shmuel N., *The Political Systems of Empires* (New York: The Free Press, 1963).

Eisenstadt, Shmuel N., 'Multiple Modernities', *Daedalus* 129:1 (2000), 1–29.

Elkins, James, 'Style', in *Grove Art Online: Oxford Art Online*, www.oxfordartonline.com/subscriber/article/grove/art/T082129, retrieved 19 August 2017.

Elliott, John H., *Empires of the Atlantic World: Britain and Spain in America 1492–1830* (New Haven, CT: Yale University Press, 2007).

Elsner, Jas, 'Style', in Robert S. Nelson and Richard Shiff (eds), *Critical Terms for Art History*, 2nd ed. (Chicago, IL/London: University of Chicago Press, 2003), pp. 98–109.

Eltis, David, et al. (eds), *The Trans-Atlantic Slave Trade: A Database*, http://slavevoyages.org, retrieved 17 May 2017.

Exquemelin, Alexander, *Histoire des avanturiers qui se sont signalez dans les Indes, contenant ce qu'ils ont fait de plus remarquable depuis vingt années* (Paris: Jacques Le Febvre, 1686; 2nd ed. Paris: Jacques Le Febvre, 1699).

Ferguson, Niall, *Empire: The Rise and Demise of the British World Order and the Lessons for Global Power* (London: Allan Lane, 2002).

Findlen, Paula, *Possessing Nature: Museums, Collecting, and Scientific Culture in Early Modern Italy* (Berkeley, CA/Los Angeles: University of California Press, 1994).

Findlen, Paula, 'Introduction: Early Modern Things: Objects in Motion, 1500–1800', in *Early Modern Things: Objects and Their Histories, 1500–1800* (London: Routledge, 2013), pp. 3–29.

Finley, Moses, 'Empire in the Greco-Roman World', *Greece & Rome* 25:1 (1978), 1–15.

Flohic, Jean-Luc, *Le patrimoine des communes de la Martinique*, 2nd ed. (Paris: Attique Editions, 2016).

Folkers, Andreas, 'Was ist neu am neuen Materialismus? Von der Praxis zum Ereignis', in Tobias Goll, Daniel Keil, and Thomas Telios (eds), *Critical Matter: Diskussionen eines neuen Materialismus* (Münster: Edition Assemblage, 2013), pp. 17–35.

Foucault, Michel, 'Of Other Spaces', *Diacritics* 16 (1986), 22–7.

Fournier, Georges, *Hydrographie contenant la théorie et la pratique de toutes les parties de la navigation* (Paris: Michel Soly, 1643).

Frank, Andre Gunder, *ReOrient: Global Economy in the Asian Age* (Berkeley, CA: University of California Press, 1998).

Freedberg, David, and Jan de Vries (eds), *Art in History, History in Art: Studies in Seventeenth-Century Dutch Culture* (Los Angeles: The Getty Center, 1991).

Freist, Dagmar, *Absolutismus* (Darmstadt: WBG, 2008).

Froger, Francois, *Relation d'un voyage fait en 1695, 1696 et 1697 aux côtes d'Afrique, détroit de Magellan, Brésil, Cayenne et isles Antilles, par une escadre des vaisseaux du roy, commandée par M. De Gennes* (Paris: de Fer, 1698).

Froger, Francois, *Relation du premier voyage des François à la Chine fait en 1698, 1699 et 1700 sur le vaisseau L'Amphitrite*, edited by E. A. Voretzsch (Leipzig: Asia Major, 1926).

Fumagalli, Maria Cristina, *On the Edge: Writing the Border between Haiti and the Dominican Republic* (Liverpool: Liverpool University Press, 2015).

Galison, Peter, *Einstein's Clocks and Poincaré's Maps: The Empire of Time* (New York: W. W. Norton, 2003).

Gallagher, John, and Ronald Robinson, 'The Imperialism of Free Trade', *Economic History Review* 6:1 (1953), 1–15.

Games, Alison, *The Web of Empire: English Cosmopolitans in an Age of Expansion, 1560–1660* (Oxford: Oxford University Press, 2008).

Gammerl, Benno, 'Emotional Styles – Concepts and Challenges', *Rethinking History: The Journal of Theory and Practice* 16:2 (2012), 161–75.

Geoffrey de Villeneuve, René Claude, *Illustrations de l'Afrique ou histoire, moeurs, usages et coutûmes des Africains. Le Sénégal* (Paris: Nepveu, 1814).

Gerbino, Anthony, *François Blondel: Architecture, Erudition, and the Scientific Revolution* (London/New York: Routledge, 2010).

Gerritsen, Anne, and Giorgio Riello, 'Spaces of Global Interactions: The Material Landscapes of Global History', in Anne Gerritsen and Giorgio Riello (eds), *Writing Material Cultural History* (London: Bloomsbury, 2014), pp. 111–35.

Gilman, Nils, *Mandarins of the Future: Modernization Theory in the Cold War* (Baltimore, MD: Johns Hopkins University Press, 2003).

Gilroy, Paul, *The Black Atlantic: Modernity and Double Consciousness*, 3rd ed. (London: Verso, 1999).

Ginzburg, Carlo, *Miti emblemi spie: morfologia e storia* (Turin: Einaudi, 1986); published in English as *Clues, Myths, and the Historical Method*, translated by John Tedeschi and Anne Tedeschi (Baltimore, MD: Johns Hopkins University Press, 1989).

Girard, Philippe, *Toussaint Louverture: A Revolutionary Life* (New York: Basic Books, 2016).

Gödel, Kurt, 'A Remark about the Relationship between Relativity Theory and Idealistic Philosophy', in Paul Arthur Schilpp (ed.), *Albert Einstein: Philosopher-Scientist* (New York: Tudor, 1951), pp. 557–62.

Godlewska, Anne, *Geography Unbound: French Geographic Science from Cassini to Humboldt* (Chicago, IL: University of Chicago Press, 1999).

Godlewska, Anne, and Neil Smith (eds), *Geography and Empire*, Institute of British Geographers Special Publication (London: Blackwell, 1994).

Goldman, Nicolas, *La nouvelle fortification* (Leiden: Elzevier, 1645).

Gould, Eliga H., *The Persistence of Empire: British Political Culture and National Identity, 1714–1783* (Chapel Hill, NC: University of North Carolina Press, 2000).

Gould, Eliga H., 'Zones of Law, Zones of Violence: The Legal Geography of the British Atlantic, circa 1772', *William and Mary Quarterly* 60:3 (2003), 471–510.

Gould, Eliga H., 'Entangled Histories, Entangled Worlds: The English-Speaking Atlantic as a Spanish Periphery', *American Historical Review* 112 (2007), 764–89.

Green, Toby, 'The Emergence of a Mixed Society in Cape Verde in the Seventeenth Century', in Toby Green (ed.), *Brokers of Change: Atlantic Commerce and Cultures in Precolonial Western Africa*, Proceedings of the British Academy, 178 (Oxford: Oxford University Press, 2012), pp. 217–36.

Grendi, Edoardo, 'Micro-analisi e storia sociale', *Quaderni storici* 35 (1977), 506–20.

Habermas, Jürgen, *Knowledge and Human Interest*, translated by Jeremy J. Shapiro (London: Heinemann, [1968] 1972).

Habermas, Jürgen, *The Theory of Communicative Action*, 2 vols, translated by Thomas McCarthy (Boston, TX: Beacon Press, [1981] 1984–9).

Hall, Stuart, 'Créolité and the Process of Creolization', in Robin Cohen and Paola Toninato (eds), *The Creolization Reader: Studies in Mixed Identities and Cultures* (Abingdon: Routledge, 2010), pp. 26–38.

Hanotaux, Gabriel, and Alfred Martineau, *Histoire des colonies françaises et de l'expansion de la France dans le monde*, 6 vols (Paris: Plon, 1930–3).

Haraway, Donna, 'Situated Knowledges: The Science Question in Feminism and the Privilege Partial Perspective', *Feminist Studies* 14:3 (1988), 575–99.

Hardt, Michael, and Antonio Negri, *Empire* (Cambridge, MA: Harvard University Press, 2000).

Hardt, Michael, and Antonio Negri, *Multitude: War and Democracy in the Age of Empire* (London: Penguin, 2004).

Hardt, Michael, and Antonio Negri, *Commonwealth* (Cambridge, MA: Harvard University Press, 2009).

Harris, Jonathan Gil, 'The New New Historicism's Wunderkammer of Objects', *European Journal of English Studies* 4 (2000), 111–23.

Harvey, David, *Social Justice and the City* (London: Arnold, 1973).

Haudrère, Philippe, *La Bourdonnais: Marin et aventurier* (Paris: Desjonquères Editions, 2013).

Haudrère, Philippe, *Les Français dans l'océan Indien (XVIIe–XIXe siècle)* (Rennes: Presses universitaire de Rennes, 2014).

Haudrère, Philippe, and Gérard Le Bouëdec, *Les Compagnies des Indes. XVIIe-XVIIIe siècles* (Rennes: Presses universitaires de Rennes, 2011).

Havard, Gilles, *Empire et métissages: Indiens et Français dans le Pays d'en Haut, 1660–1715* (Lille: Septentrion, 2003).

Hayot, E., 'Les gens de couleur libres du Fort-Royal, 1679–1823', *Revue française d'histoire d'outre-mer* 56:202/203 (1969), 5–98; 99–163.

Headrick, Daniel R., *Tools of Empire: Technology and European Imperialism in the Nineteenth Century* (New York/Oxford: Oxford University Press, 1981).

Headrick, Daniel R., *When Information Came of Age: Technologies of Knowledge in the Age of Reason and Revolution, 1700–1850* (Oxford: Oxford University Press, 2000).

Heidegger, Martin, *Sein und Zeit* (Tübingen: Max Niemeyer, 1927).

Henshall, Nicholas, *The Myth of Absolutism: Change and Continuity in Early Modern European History* (London/New York: Longman, 1992).

Hicks, Dan, and Mary C. Beaudry (eds), *The Oxford Handbook of Material Culture Studies* (Oxford: Oxford University Press, 2010).

Hinchman, Mark, 'African Rococo: House and Portrait in Eighteenth Century Senegal' (PhD Dissertation, University of Chicago, 2000); published as *Portrait of an Island: The Architecture and Material Culture of Gorée, Sénégal, 1758–1837* (Omaha, NE: University of Nebraska Press, 2015).

Hinderaker, Eric, *Elusive Empires: Constructing Colonialism in the Ohio Valley, 1673–1800* (Cambridge: Cambridge University Press, 2008).

Hinderaker, Eric, and Peter C. Mancall (eds), *At the Edge of Empire: The Backcountry in British North America* (Baltimore, MD: Johns Hopkins University Press, 2003).

Hirschman, Albert O., *The Passions and the Interests: Political Arguments for Capitalism before Its Triumph* (Princeton, NJ: Princeton University Press, 1977).

Huetz de Lemps, Christian, 'Indentured Servants Bound for the French Antilles in the Seventeenth and Eighteenth Century', in Ida Altman and James Horn (eds), *"To Make America": European Emigration in the Early Modern Period* (Berkeley, CA/Los Angeles: University of California Press, 1991), pp. 172–203.

Huyghues-Belrose, Vincent, 'Le nom des lieux à la Martinique: un patrimoine identitaire menacé', *Études caribéennes* 11 (2008), 2–17.

Inikori, Joseph E., and Stanley L. Engerman, *The Atlantic Slave Trade: Effects on Economics, Socities, and Peoples in Africa, the Americas, and Europe* (Durham, NC: Duke University Press, 1992).

Jackson, Steven J., 'Rethinking Repair', in Tarleton Gillespie, Pablo J. Boczkowski, and Kirsten A. Foot (eds), *Media Technologies: Essays on Communication, Materiality, and Society* (Cambridge, MA: The MIT Press, 2014), pp. 221–41.

James, C. L. R., *Black Jacobins: Toussaint L'Ouverture and the San Domingo Revolt* (New York: Vintage Books, [1938] 1989).

Jardine, Lisa, *Worldly Goods: A New History of the Renaissance* (London: Macmillan, 1996).

Joffe, Josef, *Uberpower: The Imperial Temptation of America* (New York: W. W. Norton, 2006).

Johnson, Chalmers, *The Costs and Consequences of the American Empire* (New York: Henry Holt, 2000).

Johnson, Chalmers, *The Sorrows of Empire: Militarism, Secrecy, and the End of the Republic* (New York: Metropolitan, 2003).

Johnston, A. J. B., 'The People of Eighteenth-Century Louisbourg', *Nova Scotia Historical Review* 11:2 (1991), 75–83.

Johnston, A. J. B., 'The Fishermen of Eighteenth-Century Cape Breton', in Eric Krause, Carol Corbin, and William O'Shea (eds), *Aspects of Louisbourg: Essays on the History of an Eighteenth-Century French Community in North America* (Sydney, Nova Scotia: University College of Cape Breton Press, 1995), pp. 198–208.

Johnston, A. J. B., *Life and Religion at Louisbourg, 1713–1758* (Montreal and Kingston/London: McGill-Queen's University Press, 1996).

Johnston, A. J. B., *Control and Order in French Colonial Louisbourg, 1713–1758* (East Lansing, MI: Michigan State University Press, 2001).

Johnston, A. J. B., *Endgame 1758: The Promise, the Glory, and the Despair of Louisbourg's First Decade* (Lincoln, NA/London: University of Nebraska Press, 2007).

Jones, Adam, *Afrika bis 1850*, Neue Fischer Weltgeschichte, 19 (Frankfurt am Main: Fischer, 2016).

Jones, Hilary, *The Métis of Senegal: Urban Life and Politics in French West Africa* (Bloomington, IN: Indiana University Press, 2013).

Joyce, Rosemary, amd Joshua Pollard, 'Archaeological Assemblages and Practices of Deposition', in Dan Hicks and Mary C. Beaudry (eds), *The Oxford Handbook of Material Culture Studies* (Oxford: Oxford University Press, 2010), pp. 291–310.

Kaeppelin, Paul, *La compagnie des Indes Orientales et François Martin. Etude sur l'histoire du commerce et des etablissements français dans l'Inde sous Louis XIV (1664–1719)* (Paris: A. Challamel, 1908; repr. New York, 1967).

Kant, Immanuel, *Perpetual Peace: A Philosophical Essay*, translated by M. Campbell Smith (London: Goerge Allen, [1795] 1911).

Kircher, Athanasius, *Mundus subterraneus in XII libros digestus* (Amsterdam: J. Janssonius/E. Weyerstraten, 1665).

Kissoun, Bruno, 'Fortifications des îles. Trois siècles d'architecture militaire en Guadeloupe: XVIIe–XIXe siècle', *Bulletin Monumental* 163:4 (2005), 343–56.

Kittsteiner, Heinz-Dieter, 'Empire. Zu den revolutionären Phantasien von Antonoio Negri und Michael Hardt', in Richard Faber (ed.), *Imperialismus in Geschichte und Gegenwart* (Würzburg: Königshausen und Neumann, 2005), pp. 125–50.

Konvitz, Josef W., *Cartography in France, 1660–1848: Science, Engineering and Statecraft* (Chicago, IL: University of Chicago Press, 1985).

Kopytoff, Igor, 'Cultural Biographies of Things: Commodization as Process', in Arjun Appadurai (ed.), *The Social Life of Things: Commodities in Cultural Perspective* (Cambridge: Cambridge University Press, 1986), pp. 64–91.

Krebs, Christopher, '"Imaginary Geography" in Caesar's *Bellum Gallicum*', *American Journal of Philology* 127 (2006), 111–36.

van Laak, Dirk, *Weiße Elefanten: Anspruch und Scheitern technischer Großprojekte im 20. Jahrhundert* (Stuttgart: DVA, 1999).

Labat, Père Jean-Baptiste, *Nouveau Voyage aux Isles de l'Amérique*, 6 vols (Paris: Guillaume Cavelier, 1722).

Labat, Jean-Baptiste, *Nouveau Voyage aux Isles de l'Amérique*, 2 vols (Den Haag: Husson, van Duren et al., 1724).

Labat, Jean-Baptiste, *Nouvelle Relation de l'Afrique Occidentale contenant une description exacte du Senegal & des Pais situés entre le Cap Blanc & la Rivière de Serreleone, jusqu'à plus de 300. lieues en avant dans les Terres. L'Histoire naturelle de ces Pais, les differentes Nations qui y sont répandues, leurs Religions & leurs moeurs. Avec l'etat ancien et present des Compagnies qui y font le Commerce. Ouvrage enrichi de Quantité de cartes, de Plans, & de Figures en taille-douce*, 5 vols (Paris: Guillaume Cavelier, 1728).

de La Courbe, Michel Jajolet, *Premier voyage du Sr. de La Courbe fait à coste d'Affrique en 1685*, edited by Prosper Cultru (Paris: Champion/Larose, 1913).

Lafleur, Gérard, *Les caraïbes des petites antilles* (Paris: Kathala, 1992).

Lamiral, Dominique Harcourt, *Illustrations de l'Afrique et du peuble africain considérés sous tous leurs rapports avec notre commerce et de nos colonies. De l'abus des privilèges exclusifs et notamment de celui de la Compagnie du Sénégal. Ce que c'est qu'une société se qualifiant d'Amis de noirs* (Paris: Dessenne, 1789).

Langlois, Gilles-Antoine, *Des villes pour la Louisiane française, théorie et pratique de l'urbanistique coloniale au XVIIIe siècle* (Paris: L'Harmattan, 2003).

de La Roncière, Charles Bourel, 'Origines du service hydrographique de la marine', *Bulletin de la Section de Géographie* 31 (1916), 6–28.

Latour, Bruno, *Science in Action: How to Follow Scientists and Engineers through Society* (Cambridge, MA: Harvard University Press, 1987).

Latour, Bruno, 'On Interobjectivity', *Mind, Culture, and Activity* 3:4 (1996), 228–45.

Latour, Bruno, *Reassembling the Social: An Introduction into Actor-Network-Theory* (Oxford: Oxford University Press, 2005).

Lavin, Irving, 'Introduction', in *Erwin Panofsky, Three Essays on Style,* edited by Irving Lavin (Cambridge, MA: The MIT Press, 1995), pp. 2–17.

Lawrence, A. W., *Trade Castles and Forts of West Africa* (London: Jonathan Cape, 1963).

Lefebvre, Henri, *La production de l'espace* (Paris: Gallimard, 1974); published in English as *The Production of Space,* translated by Donald Nicholson-Smith (Malden, MA/Oxford: Blackwell, 1991).

Leibniz, Gottfried Wilhelm, and Samuel Clarke, A *Collection of Papers Which Passed the Late Learned Mr Leibnitz and Dr Clarke, in the Years 1715 and 1716, Relating to the Prinicples of Natural Philosophy and Religion* (London: Knapton, 1717).

Lestringant, Frank, *Mapping the Renaissance World: The Geographical Imagination in the Age of Discovery* (Cambridge: Polity, 1994).

Levi, Giovanni, *L'eredità immateriale: carriera di un esorcista nel Piemonte del Seicento* (Turin: Einaudi, 1985).

Liger, Louis, *Œconomie générale de la campagne, ou La nouvelle maison rustique,* 2 vols (Paris: Charles de Sercy, 1700).

Livingstone, David N., *Putting Science in Its Place: Geographies of Scientific Knowledge* (Chicago, IL/London: University of Chicago Press, 2003).

Long, Pamela O., *Artisans/Practitioners and the Rise of New Sciences 1400–1600* (Corvallis, OR: Oregon State University Press, 2011).

Ly, Abdoulaye, *La Compagnie du Sénégal* (Paris: Présence africaine, 1958).

Lynch, Andrew, 'Emotional Community', in Susan Broomhall (ed.), *Early Modern Emotions: An Introduction* (London/New York: Routledge, 2017), pp. 3–7.

Mahoney, Michael S., 'Organizing Expertise: Engineering and Public Works under Jean-Baptiste Colbert, 1662–83', *Osiris* 25, Special Issue: Expertise: Practical Knowledge and the Early Modern State, edited by Eric H. Ash (2010), 149–70.

Malleson, George Bruce, *History of the French in India: From the Founding of Pondichery in 1674 to the Capture of That Place in 1761,* 2nd ed. (Edinburgh: John Grant, 1909).

Mandler, Peter, and Deborah Cohen, 'The History Manifesto – A Critique', *The American Historical Review* 120:2 (2015), 530–42.

Mann, Emily, 'To Build and Fortify: Defensive Architecture in the Early Atlantic Colonies', in Daniel Maudlin and Bernard L. Herman (eds), *Building the British Atlantic World: Spaces, Places and Material Culture, 1600–1850* (Chapel Hill, NC: University of North Carolina Press, 2016), pp. 31–52.

Mann, Emily, 'Viewed from a Distance: Eighteenth-Century Images of Fortifications on the Coast of West Africa', in Kohn Kwadwo Osei-Tutu and Victoria Smith (eds), *Shadows of Empire in West Africa: New Perspectives on European Fortifications* (London: Palgrave Macmillan, 2017), pp. 107–36.

Mann, Michael, *The Incoherent Empire* (New York: Verso, 2002).

Marais, Henri, *Introduction géométrique à l'étude de la relativité* (Paris: Gauthier-Villars, 1923)

Maran, René, *Les Pionniers de l'Empire*, 3 vols (Paris: A. Michel, 1943–55).

Marcocci, Giuseppe, 'Too Much to Rule: States and Empires across the Early Modern World', *Journal of Early Modern History* 20 (2016), 511–25.

Mark, Peter, *"Portuguese" Style and Luso-African Identity: Precolonial Senegambia, Sixteenth–Nineteenth Centuries* (Bloomington, IN: Indiana University Press, 2002).

Marquet, Julie, 'Le rôle des intermédiaires dans l'implantation colonial française. L'exemple de la famille de Tiruvengadam à Pondichéry au XVIIIe siècle', *Encyclo. Revue de l'école doctorale ED 382*, Université Sorbonne Paris Cité (2014), 17–42.

Marshall, Bill, *The French Atlantic: Travels in Culture and History* (Liverpool: Liverpool University Press, 2009).

Martin, Rex, and David Reidy (eds), *Rawls's Law of Peoples: A Realistic Utopia?* (Malden, MA/Oxford: Blackwell, 2006).

Martineau, Alfred, *Dupleix et l'Inde Française*, 4 vols (Paris: Champion, 1920–8).

Marx, Karl, *Das Kapital* (Berlin: Dietz, 1962).

Mauny, Raymond, 'Les prétendues navigations dieppoises à la Côte occidentales d'Afrique au XIVe siècle', *Bulletin de l'Institut Fondamental de l'Afrique Noire* 12 B:1 (1950), 122–34.

May, L.-P., *Histoire économique de la Martinique (1635–1763)*, Thèse pour le doctorat de droit (Paris: Les Presses Modernes, 1930).

McAleer, John, 'Objects of Empire: Museums, Material Culture and Histories of Empire', in Anne Gerritsen and Giorgio Riello (eds), *Writing Material Cultural History* (London: Bloomsbury, 2014), pp. 249–57.

McCabe, Ina Baghdiantz, *Orientalism in Early Modern France: Eurasian Trade, Exoticism and the Ancien Régime* (Oxford: Berg, 2008).

McClellan, James E., *Colonialism and Science: Saint Domingue in the Old Regime* (Baltimore, MD/London: Johns Hopkins University Press, 1992).

McClellan, James E., and François Regourd, *The Colonial Machine: French Science and Overseas Expansion in the Old Regime* (Turnhout: Brepols, 2010).

McLennan, J. S., *Louisbourg from Its Foundation to Its Fall* (1918; repr. Halifax: Book Room, 1983).

McNeill, John Robert, *Atlantic Empires of France and Spain: Louisbourg and Havana, 1700–1763* (Chapel Hill, NC: University of North Carolina Press, 1985).

McNeill, William H., *The Rise of the West: A History of the Human Community* (Chicago, IL: University of Chicago Press, 1963).

Meli, Domenico Bertoloni, *Thinking with Objects: The Transformation of Mechanics in the Seventeenth Century* (Baltimore, MD: Johns Hopkins University Press, 2006).

Meyers, Fred (ed.), *The Empire of Things: Regimes of Value and Material Culture* (Santa Fe, NM/Oxford: School of American Research Press/James Currey, 2001).

Mignolo, Walter D., *The Darker Side of Renaissance: Literacy, Territoriality, and Colonization* (Ann Arbor, MI: University of Michigan Press, 1995).

Miller, Daniel, *Material Culture: Why Some Things Matter* (Chicago, IL: University of Chicago Press, 1998).

Miller, Daniel, *Materiality* (Durham, NC: Duke University Press, 2005).

Miller, Daniel, *Stuff* (Cambridge: Polity, 2010).

Miller, Dayton C., 'The Ether-Drift Experiment and the Determination of the Absolute Motion of the Earth', *Review of Modern Physics* 5 (1933), 203–42.

Mintz, Sidney W., *Sweetness and Power: The Place of Sugar in Modern History* (New York: Viking, 1985).

Mißfelder, Jan-Friedrich, *Das Andere der Monarchie. La Rochelle und die Idee der "monarchie absolue" in Frankreich, 1568–1630*, Pariser Historische Studien, 97 (München: Oldenbourg, 2012).

del Monte y Tejada, Antonio, *Historia de Santo Domingo* (Santo Domingo: Garcia Herman, 1890).

Moon, Donald, *John Rawls: Liberalism and the Challenges of Late Modernity* (London: Rowman & Littlefield, 2014).

Moreau de Saint-Méry, Médéric-Louis-Elle, *Loix et constitutions des colonies françoises de l'Amérique sous le vent*, 6 vols (Paris: Moutard et al., 1784–90).

Moreau de Saint-Méry, Médéric Louis Elle, *Recueil de vues des lieux principaux de la colonie françoise de Saint-Domingue, gravée par les soins de M. Ponce [...] accompagné de Cartes et Plans de la même Colonie, gravés par les soins de M. Phelipeau, Ingénieur-Géographe* (Paris: Moreau de Saint-Méry/ Ponce/Phelipeau, 1791).

Moreau de Saint-Méry, Médéric Louis Elle, *Description topographique, physique, civile, politique et historique de la partie française de Saint-Domingue; avec des observations générales sur la population, accompagnées de détails les plus propres à faire connaître l'état de la colonie à l'époque du 18 octobre 1789* (Philadelphia, PA: Chez l'auteur, 1797–8).

Moss, William, 'The Archaeology of a North American and the Early Modern Period in Québec', *Post-Medieval Archaeology* 43:1, Special Issue: The Recent Archaeology of the Early Modern Period in Québec City (2009), 1–12.

Moyn, Samuel, and Andrew Sartori, 'Approaches to Global Intellectual History', in Samuel Moyn and Andrew Sartori (eds), *Global Intellectual History* (New York: Columbia University Press, 2013), pp. 3–30.

Münkler, Herfried, *Imperien. Die Logik der Weltherrschaft* (Berlin: Rowohlt, 2005).

Mukerji, Chandra, *From Graven Images: Patterns of Modern Materialism* (New York: Columbia University Press, 1983).

Mukerji, Chandra, *Territorial Ambitions and the Gardens of Versailles* (Cambridge: Cambridge University Press, 1997).

Mukerji, Chandra, 'Cartography, Entrepreneurialism, and Power in the Reign of Louis XIV: The Case of the Canal du Midi', in Pamela H. Smith and Paula Findlen (eds), *Merchants and Marvels: Commerce, Science, and Art in Early Modern Europe* (New York: Routledge, 2002), pp. 248–76.

Mukerji, Chandra, 'Stewardship Politics and the Control of Wild Weather: Levees, Seawalls, and State Building in 17th-Century France', *Social Studies of Science* 37:1 (2007), 127–33.

Mukerji, Chandra, *Impossible Engineering: Technology and Territoriality on the Canal du Midi* (Princeton, NJ: Princeton University Press, 2009).

Navarro-Andraud, Zélie, 'Les élites urbaines de Saint-Domingue dans la seconde moitié du 18e siècle: La place des administrateurs coloniaux (1763–1792)' (PhD Dissertation, Université de Toulouse II-Le Mirail, 2007).

Navarro-Andraud, Zélie, 'La résidence urbaine des administrateurs coloniaux de Saint-Domingue dans la seconde moitié du XVIIIe siècle', *Articulo. Journal of Urban Research*, Special Issue 1: Occupying, Organising and Ordering Urban Space (2009), http://articulo.revues.org/997, retrieved 7 June 2017.

Newitt, Malyn D. D., *A History of Portuguese Overseas Expansion, 1400–1668* (London: Routledge, 2005).

Newton, Isaac, *Philosophiae naturalis principia mathematica* (London: Innys, 1726).

Nimako, Kwame, and Glenn Willemsen (eds), *The Dutch Atlantic: Slavery, Abolition and Emancipation* (London: Pluto Press, 2011).

Nolte, Hans-Heinrich, *Weltgeschichte. Imperien, Religionen und Systeme, 15.–19. Jahrhundert* (Wien/Köln/Weimar: Böhlau, 2005).

Nordman, Daniel, *Frontières de France. De l'espace au territoire, XVIe–XIXe siècle* (Paris: Gallimard, 1998).

Nussbaum, Martha, 'Kant and Stoic Cosmopolitanism', *The Journal of Political Philosophy* 5:1 (1997), 1–26.

van Oers, Ron, *Dutch Town Planning Overseas During VOC and WIC Rule* (Zutphen: Walburg, 2000).

O'Malley, Michelle, and Evelyn Welch, *The Material Renaissance* (Manchester: Manchester University Press, 2007).

d'Orgeix, Emilie, and Céline Frémaux, 'La petite maison dans les abattis ou l'art de rédiger aux bois par Jean Antoine de Brûletout, chevalier de Préfontaine dans son habitation de la France équinoxiale (1754–1763)', *In Situ. Revue des patrimoines* 21 (2013), http://insitu.revues.org/10338, retrieved 23 May 2017.

Osterhammel, Jürgen, *Die Verwandlung der Welt. Eine Geschichte des 19. Jahrhunderts*, 3rd ed. (München: C. H. Beck, 2009).

Padrón, Ricardo, *The Spacious Word: Cartography, Literature, and Empire in Early Modern Spain* (Chicago, IL: University of Chicago Press, 2004).

Parry, J. H., *The Spanish Seaborn Empire* (Berkeley, CA/Los Angeles/London: University of California Press, [1966] 1990).

Pérotin-Dumon, Anne, *La ville aux Iles, la ville dans l'île: Basse-Terre et Pointe-à-Pitre, 1650–1820* (Paris: Karthala, 2000).

Perry, Curtis, *Material Culture and Cultural Materialisms in the Middle Ages and Renaissance* (Turnhout: Brepols, 2001).

Pestana, Carla Gardina, *Protestant Empire: Religion and the Making of the British Atlantic World* (Philadelphia, PA: University of Pennsylvania Press, 2010).

Petitjean-Roget, Jacques, *Le Gaoulé: La révolte de la Martinique, 1717* (Fort-de-France: Société de l'histoire de la Martinique, 1966).

Petitjean-Roget, Jacques, *Les premières habitations de la Martinique. Monuments historiques, Architecture d'outremer* (Paris: Editions CNMHS, 1981).

Pickering, Andrew, *The Mangle of Practice* (Chicago, IL: University of Chicago Press, 1995).

Pillai, Ananda Ranga, *The Diary of Ananda Ranga Pillai*, 12 vols, translated from the Tamil by order of the Government of Madras, edited by H. Dodwell (Madras: Government Press, 1922).

Pitts, Jennifer, 'Political Theory of Empire and Imperialism', *Annual Review of Political Science* 13 (2010), 211–35.

Pluchon, Pierre, *Histoire des Antilles et de la Guyane* (Toulouse: Privat, 1982).

Pluchon, Pierre, *Histoire de la colonisation française*, vol. 1: Le Premier empire colonial. Des origines à la Restauration (Paris: Fayard, 1991).

Pollock, Sheldon, 'Empire and Imitation', in Craig J. Calhoun, Frederick Cooper, and Kevin W. Moore (eds), *Lessons of Empire: Imperial Histories and American Power* (New York: New Press, 2006), pp. 175–88.

Pomeranz, Kenneth, *The Great Divergence: China, Europe, and the Making of the Modern World Economy* (Princeton, NJ: Princeton University Press, 2000).

Popkin, Jeremy D., *A Concise History of the Haitian Revolution* (Malden, MA/Oxford: Wiley-Blackwell, 2012).

Pratt, Mary Louise, *Imperial Eyes: Travel Writing and Transculturation* (London: Routledge, 1992).

Préfontaine, Jean Antoine de Brûletout de, *Maison rustique, à l'usage de la partie de la France équinoxiale connue sous le nom de Cayenne* (Paris: J. B. Bauche, 1763).

Priestley, Herbert Ingram, *France Overseas through the Old Régime: A Study of European Expansion* (New York/London: Appleton-Century, 1939).

Pritchard, James, *In Search of Empire: The French in the Americas, 1670–1730* (Cambridge: Cambridge University Press, 2004).

de Provins, Pacifique, *Brieve Relation du voyage des Isles de l'Amérique* (Paris: Nicolas et Jean de la Coste, 1646).

Raj, Kapil, *Relocating Modern Science: Circulation and Construction of Knowledge in South Asia and Europe, 1650–1900* (Basingstoke: Palgrave Macmillan, 2008).

Rau, Susanne, *Räume. Konzepte, Wahrnehmungen, Nutzungen* (Frankfurt am Main: Campus, 2013).

Rau, Susanne, and Benjamin Steiner, 'Europäische Grenzordnungen in der Welt. Ein Beitrag zur Historischen Epistemologie der Globalgeschichtsschreibung', *Themenportal Europäische Geschichte* (2013), www.europa.clio-online.de/2013/Article=611, retrieved 7 March 2017.

Rau, Susanne, and Benjamin Steiner, 'Raumforschung, historische', in Friedrich Jaeger (ed.), *Enzyklopädie der Neuzeit* (2017), http://dx.doi.org/10.1163/2352-0248_edn_a6014000, retrieved 7 March 2017.

Rawls, John, 'The Law of Peoples', *Critical Inquiry* 20:1 (1991), 36–68 (exp. ed. Cambridge, MA: Harvard University Press, 1999).

Rawls, John, *A Theory of Justice*, revised ed. (Cambridge, MA: Belknap, [1971] 1999).

Reckwitz, Andreas, 'Affective Spaces: A Praxeological Outlook', *Rethinking History: The Journal of Theory and Practice* 16:2 (2012), 241–58.

Reddy, William M., 'Against Constructionism: The Historical Ethnography of Emotions', *Current Anthropology* 38 (1997), 327–51.

Reddy, William M., *The Navigation of Feeling* (Cambridge: Cambridge University Press, 2001).

Rensmann, Lars, 'Back to Kant? The Democratic Deficits in Habermas' Global Constitutionalism', in Tom Bailey (ed.), *Deprovincializing Habermas: Global Perspectives* (London: Routledge, 2015), pp. 27–50.

Riello, Giorgio, 'Things That Shape History: Material Culture and Historical Narratives', in Karen Harvey (ed.), *History and Material Culture* (London: Routledge, 2009), pp. 24–46.

Ritter, Karl, *Naturhistorische Reise nach der westindischen Insel Hayti* (Stuttgart: Hallberger, 1836).

Roberts, Lissa, Simon Schaffer, and Peter Dear (eds), *The Mindful Hand: Inquiry and Invention from Late Renaissance to Early Industrialization* (Amsterdam: Koninklijke Nederlandse Akademie van Wetenschappen, 2007).

Robertson, Roland, *Globalization: Social Theory and Global Culture* (London: Sage, 1992).

Rochas d'Aiglun, Albert, *Vauban. Sa famille et ses écrits, ses oisivetés et sa correspondance, analyse et extraits*, 2 vols (Geneva: Slatkine, [1910] 1972).

de Rochefort, Charles, *Histoire naturelle et morale des iles Antilles de l'Amérique. Enrichie de plusieurs belles figures des raretez qui y sont décrites. Avec un vocabulaire caraïbe* (Rotterdam: Arnould Leers, 1658).

Rodney, Walter, *How Europe Underdeveloped Africa* (London: Bogle-L'Ouverture Publications, 1972).

Rosenwein, Barbara H., *Emotional Communities in the Early Middle Ages* (Ithaca, NY: Cornell University Press, 2006).

Rosenwein, Barbara H., 'Problems and Methods in the History of Emotions', *Passions in Context: Journal of the History and Philosophy of Emotions* 1:1 (2010), 12–24.

Rothschild, Emma, *The Inner Life of Empires: An Eighteenth-Century History* (Princeton, NJ: Princeton University Press, 2011).

Roulet, Eric, *La Compagnie des Iles de l'Amérique (1635–1651): Une entreprise colonial au XVIIe siècle* (Rennes: Presses universitaires de Rennes, 2017).

Roussier, Paul (ed.), *L'établissement d'Issiny, 1687–1702* (Paris: Larose, 1935).

Russell-Wood, Anthony J. R., *The Portuguese Empire, 1415–1808: A World on the Move* (Baltimore, MD: John Hopkins University Press, 1998).

Russell-Wood, Anthony J. R., 'The Portuguese Atlantic, 1415–1808', in Jack P. Greene and Philip D. Morgan (eds), *Atlantic History: A Critical Appraisal* (Oxford: Oxford University Press, 2009), pp. 81–111.

Sainton, Jean-Pierre (ed.), *Histoire et Civilisation de la Caraïbe: Gouadeloupe, Martinique, Petites Antilles. La construction des sociétés antillaises des origines au temps présent: Structures et dynamiques*, Tome 1: Le temps des Genèses; des origines à 1685 (Paris: Karthala, 2004).

Sarmant, Thierry, and Mathieu Stoll, *Régner et gouverner. Louis XIV et ses ministers* (Paris: Perrin, 2010).

Sauphanor, Maryse, *Maison creole* (Paris: Les auteur indépendantes, 2004).

Seneca, *Epistles, Volume II: Epistles 66–92*, translated by Richard Mott Gummere (Cambridge, MA: Harvard University Press, 1920).

Schäfer, Dagmar, *The Crafting of the 10,000 Things: Knowledge and Technology in Seventeenth-Century China* (Chicago, IL: University of Chicago Press, 2011).

Schaffer, Simon, James Delbourgo, Kapil Raj, and Lissa Roberts (eds), *The Brokered World: Go-Betweens and Global Intelligence, 1770–1820* (Sagamore Beach, MA: Science History Publications, 2009).

Schama, Simon, *The Embarrassment of Riches: An Interpretation of Dutch Culture in the Golden Age* (London: Collins, 1988).

Schiebinger, Londa, *Plants and Empire: Colonial Bioprospecting in the Atlantic World* (Cambridge, MA: Harvard University Press, 2004).

Schmidt, Benjamin, 'The Dutch Atlantic: From Provincialism to Globalism', in Jack P. Greene and Philip D. Morgan (eds), *Atlantic History: A Critical Appraisal* (Oxford: Oxford University Press, 2009), pp. 163–90.

Schmidt, Benjamin, *Inventing Exoticism: Geography, Globalism, and Europe's Early Modern World* (Philadelphia, PA: University of Pennsylvania Press, 2015).

Schmitt, Carl, *The Nomos of the Earth in the International Law of the Jus Publicum Europaeum* (New York: Telos Press, 2003).

Schwartz, Stuart B. (ed.), *Implicit Understandings: Observing, Reporting, and Reflecting on the Encounters between Europeans and Other People in the Early Modern Era* (Cambridge: Cambridge University Press, 1994).

Schwartz, Stuart B., *Tropical Babylons: Sugar and the Making of the Atlantic World, 1450–1680* (Chapel Hill, NC/London: University of North Carolina Press, 2004).

Scott, David, *Conscripts of Modernity: The Tragedy of Colonial Enlightenment* (Durham/London: Duke University Press, 2004).

Scott, James C., *Seeing Like a State: How Certain Schemes to Improve the Human Condition Have Failed* (New Haven, CT/London: Yale University Press, 1998).

Secord, James A., 'Knowledge in Transit', *Isis* 95:4 (2004), 654–72.

Shils, Edward, 'Centre and Periphery', in Polanyi Festschrift Committee (eds), *The Logic of Personal Knowledge: Essays in Honour of Michael Polanyi* (Glencoe, IL: Free Press, 1961), pp. 117–30.

Shils, Edward, *Center and Periphery: Essays in Macrosociology* (Chicago, IL: University of Chicago Press, 1975).

Simmel, Georg, *The Philosophy of Money* (London: Routledge, [1900] 1978).

Smith, Pamela H., *The Body of the Artisan: Art and Experience in the Scientific Revolution* (Chicago, IL: University of Chicago Press, 2006).

Smith, Pamela H., and Benjamin Schmidt (eds), *Making Knowledge in Early Modern Europe: Practices, Objects, and Texts, 1400–1800* (Chicago, IL: University of Chicago Press, 2008).

Smith, Pamela H., Amy Meyers, and Harold J. Cook (eds), *Ways of Making and Knowing: The Material Culture of Empirical Knowledge* (Ann Arbor, MI: University of Michigan Press, 2013).

Soja, Edward W., *Postmodern Geographies: The Reassertion of Space in Critical Social Theory* (London: Verso, 1989).

Soja, Edward W., *Thirdspace* (Malden, MA: Blackwell, 1996).

Soll, Jacob, *The Information Master: Jean Baptiste Colbert's Secret State Intelligence System* (Ann Arbor, MI: University of Michigan Press, 2009).

Sparks, Randy J., *The Two Princes of Calabar: An Eighteenth-Century Atlantic Odyssey* (Cambridge, MA: Harvard University Press, 2004).

Sparks, Randy J., *Where the Negroes Are Masters: An African Port in the Era of the Slave Trade* (Cambridge, MA/London: Harvard University Press, 2014).

St Clair, William, *The Grand Slave Emporium: Cape Coast Castle and the British Slave Trade* (London: Profile Books, 2007).

Steiner, Benjamin, *Colberts Afrika. Eine Wissens- und Begegnungsgeschichte in Afrika im Zeitalter Ludwigs XIV.* (München: Oldenbourg, 2014).

Steiner, Benjamin, 'The Monuments of Empire: Global Material Culture, "Colonial" Spaces and Emotional Styles in French Senegambia (c. 1630 – c. 1730)', *Cromohs* 20 (2015), 52–76.

Steiner, Benjamin, 'La première Compagnie du Sénégal de Rouen de 1633', in Éric Roulet (ed.), *Le monde des compagnies: Les premières companies dans l'Atlantique*, vol. 1: Structures et modes de fonctionnement (Aachen: Shaker, 2017), pp. 145–59.

Stoler, Ann Laura, 'Imperial Formations and the Opacities of Rule', in Craig J. Calhoun, Frederick Cooper, and Kevin W. Moore (eds), *Lessons of Empire: Imperial Histories and American Power* (New York: New Press, 2006), pp. 48–60.

Stoler, Ann Laura, and Frederick Cooper, 'Between Metropole and Colony: Rethinking a Research Agenda', in Ann Laura Stoler and Frederick Cooper (eds), *Tensions of Empire: Colonial Cultures in a Bourgeois World* (Berkeley, CA: University of California Press, 1997), pp. 1–58.

Stone, Jeffrey C., 'Imperialism, Colonialism and Cartography', *Transactions of the Institute of British Geographers* 13 (1988), 57–64.

Subrahmanyam, Sanjay, 'Connected Histories: Notes towards a Reconfiguration of Early Modern Eurasia', *Modern Asian Studies* 31:3 (1997), 735–62.

Subrahmanyam, Sanjay, 'Hearing Voices: Vignettes of Early Modernity in South Asia, 1400–1750', *Daedalus* 127 (1998), 75–104.

Subrahmanyam, Sanjay, *Explorations in Connected History*, vol. 1: From the Tagus to the Ganges; vol. 2: Mughals and Franks (Oxford: Oxford University Press, 2005).

Subrahmanyam, Sanjay, 'Between a Rock and a Hard Place', in Simon Schaffer, James Delbourgo, Kapil Raj, and Lissa Roberts (eds), *The Brokered World: Go-Betweens and Global Intelligence, 1770–1820* (Sagamore Beach, MA: Science History Publications, 2009), pp. 429–40.

Subrahmanyam, Sanjay, 'Introduction', in Jerry H. Bentley, Sanjay Subrahmanyam, and Merry E. Wiesner-Hank (eds), *The Cambridge World History*, vol. 6: The Construction of a Global World, 1400–1800 CE, 2 parts (Cambridge: Cambridge University Press, 2015), part 2, pp. xvii–xx.

de Talleyrand, Charles-Maurice, *Essai sur les avantages à retirer de colonies nouvelles dans les circonstance présentes, par le citoyen Talleyrand. Lu à la séance publique de l'Institut national le 15 méssidor an 5* [1797]; published in English as 'An Essay on the Advantages to Be Derived from New Colonies, in the Present Circumstances', *The Colonial Journal* 4 (1816), 322–8.

Tarantino, Giovanni, 'Mapping Religion (and Emotions) in the Protestant Valleys of Piedmont', *Asdiwal* 9 (2014), 91–105.

Tarrade, Jean, *Le commerce colonial de France à la fin de l'Ancien Régime: l'évolution du régime de l'exclusif de 1763 à 1789*, 2 vols (Paris: Presses universitaires de France, 1972).

Du Tertre, P. Jean-Baptiste, *Histoire générale des Isles de Saint-Christophe, de la Guadeloupe, de la Martinique* (Paris: Jacques Langlois, 1654).

Du Tertre, P. Jean-Baptiste, *Histoire générale des Antilles Habitées par les François divisée en deux tomes, Et enrichie de Cartes & de Figures*, 2 vols (Paris: Thomas Iolly, 1667).

Thiaw, Ibrahima, 'An Archaeological Investigation of Long-Term Culture Change in the Lower Falemme (Upper Senegal Region) A.D. 500–1900' (PhD Dissertation, Rice University, 1999).

Thiaw, Ibrahima, 'Every House Has a Story: The Archaeology of Gorée Island, Sénégal', in Livio Sansone, E. Soumonni, and Boubacar Barry Africa (eds), *Brazil and the Construction of Trans-Atlantic Black Identities* (Trenton, NJ: Africa World Press, 2008), pp. 45–62.

Thiaw, Ibrahima, 'L'espace entre les mots et les choses: mémoire historique et culture matérielle à Gorée (Sénégal)', in Ibrahima Thiaw (ed.), *Espace, culture matérielle et identités en Sénégambie* (Dakar: Codesria, 2010), pp. 18–41.

Thiaw, Ibrahima, 'Atlantic Impacts on Inland Senegambia: French Penetration and African Initiatives in Eighteenth- and Nineteenth-Century Gajaaga and Bundu (Upper Senegal River)', in J. Cameron Monroe and Akinwumi Ogundiran (eds), *Power and Landscape in Atlantic West Africa: Archaeological Perspectives* (Cambridge: Cambridge University Press, 2012), pp. 49–77.

Thornton, John K., *Africa and the Africans in the Making of the Atlantic World, 1400–1800*, 2nd ed. (Cambridge: Cambridge University Press, 1998).

Thornton, John K., *A Cultural History of the Atlantic World, 1250–1820* (Cambridge: Cambridge University Press, 2012).

Thornton, John K., and Linda M. Heywood, *Central Africans, Atlantic Creoles, and the Foundation of America* (Cambridge: Cambridge University Press, 2007).

Tilley, Chris, Webb Kayne, Susan Kuechler, Mike Rowlands, and Patricia Spyer (eds), *The Handbook of Material Cultures* (London: Sage, 2006).

Todd, David, 'A French Imperial Meridian, 1814–1870', *Past and Present* 210 (2011), 155–86.

Todd, Emmanuel, *La Chute finale: Essai sur la décomposition de la sphère soviétique* (Paris: Édition Robert Laffont, 1976).

Todd, Emmanuel, *Après l'empire. Essai sur la décomposition du système américaine* (Paris: Gallimard, 2002).

Trentmann, Fred, *Empire of Things: How We Became a World of Consumers, from the Fifteenth Century to the Twenty-First* (London: Allen Lane, 2016).

Trivellato, Francesca, *The Familiarity of Strangers: The Sephardic Diaspora, Livorno, and Cross-Cultural Trade in the Early Modern Period* (New Haven, CT: Yale University Press, 2009).

Trivellato, Francesca, 'Is There a Future for Italian Microhistory in the Age of Global History?', *California Italian Studies* 2 (2011), https://escholarship.org/uc/item/0z94n9hq, retrieved 22 January 2020.

Trouillot, Michel-Rolph, *Silencing the Past: Power and the Production of History* (Boston, TX: Beacon Press, 1995).

Trouillot, Michel-Rolph, 'Undenkbare Geschichte. Zur Bagatellisierung der haitischen Revolution', in Sebastian Conrad and Shalini Randeria (eds), *Jenseits des Eurozentrismus. Postkoloniale Perspektiven in den Geschichtswissenschaften* (Frankfurt am Main/New York: Campus, 2002), pp. 84–115.

Tully, James, 'The Kantian Idea of Europe: Critical and Cosmopolitan Perspectives', in Anthony Pagden (ed.), *The Idea of Europe: From Antiquity to the European Union* (Cambridge: Cambridge University Press, 2002), pp. 331–58.

Turnbull, David, 'Travelling Knowledge: Narratives, Assemblage and Encounters', in Marie-Noelle Bourguet, Christian Licoppe, and H. Otto Sibum (eds), *Instruments, Travel, and Science: Itineraries of Precision from the Seventeenth to the Twentieth Century* (London/New York: Routledge, 2002), pp. 273–94.

Turner, Frederick J., 'The Significance of the Frontier in American History', *Proceedings of the State Historical Society of Wisconsin* (1893); repr. *The Frontier in American History* (New York: Holt, 1921).

Vaghi, Massmiliano, 'Alfred Martineau et la « genèse » du protectorat. Le cas indien (1745–1761)', *French Colonial History* 14 (2013), 71–88.

de Vaissière, Pierre, *Saint-Domingue: la société et la vie créole sous l'ancien régime, 1629–1789* (Paris: Perrin, 1909).

Vallières, Marc, *Quebec City: A Brief History* (Quebec: Les Presses de l'Université Laval, 2011).

Vansina, Jan, *Oral Tradition as History* (Madison, WI: University of Wisconsin Press, 1988).

de Vastey, Pompée Valentin, *An Essay on the Causes of the Revolution and Civil Wars of Hayti, Being a Sequel to the Political Remarks upon Certain French Publications and Journals Concerning Hayti*, translated by W. H. M. B. (Exeter: Western Luminary Office, 1823).

Vergé-Franceschi, Michel, *La Marine française au XVIIIe siècle* (Paris: SEDES, 1996).

Verrand, Laurence, *La vie quotidienne des Indiens caraïbes aux Petites Antilles (XVIIe siècle)* (Paris: Karthala, 2001).

Vidal, Cécile, 'Africains et Européens au pays des Illinois durant la période française (1699–1765)', *French Colonial History* 3 (2003), 51–68.

Vidal, Laurent, and Émile D'Orgeix (eds), *Les villes françaises du nouveau monde* (Paris: Somogy, 1999).

Vigié, Marc, *Dupleix* (Paris: Fayard, 1993).

de Ville, Antoine, *Les fortifications du chevalier Antoine de Ville, contenans la manière de fortifier toute sorte de places [...] avec l'ataque et les moyens de prendre les places [...] plus la défense* (Lyon: Philippe Borde, 1640).

Voltaire, 'Précis du siècle de Louis XV', in *Œuvres historiques* (Paris: Gallimard, 1957), pp. 1454–62.

Wallerstein, Immanuel, *The Modern World-System I: Capitalist Agriculture and the Origins of the European World-Economy in the Sixteenth Century*,

Studies in Social Discontinuity (New York/San Francisco, CA/London: Academic Press, 1974).

Wallerstein, Immanuel, *The Modern World-System II: Mercantilism and the Consolidation of the European World-Economy, 1600–1750*, Studies in Social Discontinuity (New York/London/Toronto: Academic Press, 1980).

Wallerstein, Immanuel, *The Modern World System III. The Second Era of Great Expansion of the Capitalist World-Economy, 1730–1840s* (Amsterdam: Elsevier, 1989).

Weber, David J., 'Turner, the Boltonians, and the Borderlands', *The American Historical Review* 91 (1986), 66–81.

Weber, Jacques, *Compagnies et Comptoirs. L'Inde des Français, XVII–XXème siècles* (Paris: Société française d'Outre-Mer, 1991).

Weber, Jacques, *Les relations entre l'Inde et la France, de 1673 à nos jours* (Paris: Les Indes savantes, 2002).

Welch, Evelyn, *Shopping in the Renaissance: Consumer Cultures in Italy 1400–1600* (Oxford: Oxford University Press, 2005).

White, Richard, *The Middle Ground: Indians, Empires, and Republics in the Great Lakes Region, 1650–1815* (Cambridge: Cambridge University Press, 1991).

Williams, Raymond, *Marxism and Literature* (Oxford: Oxford University Press, 1977).

Wolf, Eric R., and Sidney W. Mintz, 'Haciendas and Plantations in Middle America and the Antilles', *Social and Economic Studies* 6 (1957), 380–412.

Wynne-Jones, Stephanie, *A Material Culture: Consumption and Materiality on the Coast of Precolonial East Africa* (Oxford: Oxford University Press, 2016).

Zeuske, Michael, *Sklaven und Sklaverei in den Welten des Atlantiks 1400–1940: Umrisse, Anfänge, Akteure, Vergleichsfelder und Bibliographien* (Münster: Lit Verlag, 2006).

INDEX

EU authorised representative for GPSR:
Easy Access System Europe, Mustamäe tee 50,
10621 Tallinn, Estonia
gpsr.requests@easproject.com

www.ingramcontent.com/pod-product-compliance
Lightning Source LLC
Chambersburg PA
CBHW072040280526
45788CB00006B/2133